The EU and Global Climate Justice

This book examines the European Union (EU)'s contribution to the development of the global climate regime within the broader framework of global justice.

It argues that the procedural dimension of justice has been largely overlooked so far in the assessment of EU climate policy and reveals that the EU has significantly contributed to the development of the climate regime within its broader efforts to 'solidarise' international society. At the same time, the book identifies deficits of the climate regime and limits to the EU's impact, and explains why the EU policy towards global climate change has shifted over time. Finally, it argues that these policies should not be assessed in terms of being either wholly positive or wholly negative, but that they are shot through with ambiguities.

This book will be of key interest to scholars, students, and practitioners of climate change, climate politics, and environmental and climate justice studies, and more broadly to EU Studies and International Relations.

Franziskus von Lucke is a Postdoctoral Researcher at the Institute of Political Science at the University of Tübingen, Germany.

Thomas Diez is Professor of Political Science and International Relations at the University of Tübingen, Germany.

Solveig Aamodt is a Reseacher at ARENA Centre for European Studies, University of Oslo, and a Senior Researcher at CICERO Center for International Climate Research, Oslo, Norway.

Bettina Ahrens is a Research Manager at the University of Stuttgart, Germany.

Routledge Studies on the European Union and Global Order

Series editor: Helene Sjursen, ARENA Centre for European Studies, University of Oslo, Norway.

Some critical voices notwithstanding, the scholarly debate on the EU's foreign policy role and its contribution to global values has rested on the assumption of the viability of a liberal world order with the EU at its vanguard. This series revisits this assumption.

The series analyses the external policies - and the response from external actors - of the European Union (EU) at a time of enhanced uncertainty, risk, and ambiguity. Drawing on a threefold conception of global political justice, it offers an innovative account of the EU's global role and relevance at a time of profound contestation over global norms. It delivers in-depth analyses of a set of core issues of global governance in which the EU has played a major role, amongst them migration, climate change, security and conflict, and development. Through these analyses, the Series re-conceptualises the EU's global role, and brings forth a new perspective on the crisis of the liberal world order; on what is at stake and for whom.

The EU and Global Climate Justice
Normative Power Caught in Normative Battles
Franziskus von Lucke, Thomas Diez, Solveig Aamodt and Bettina Ahrens

The EU and Global Climate Justice

Normative Power Caught in Normative Battles

Franziskus von Lucke, Thomas Diez, Solveig Aamodt and Bettina Ahrens

Routledge
Taylor & Francis Group
LONDON AND NEW YORK

First published 2021
by Routledge
2 Park Square, Milton Park, Abingdon, Oxon OX14 4RN

and by Routledge
52 Vanderbilt Avenue, New York, NY 10017

Routledge is an imprint of the Taylor & Francis Group, an informa business

British Library Cataloguing-in-Publication Data
A catalogue record for this book is available from the British Library

Library of Congress Cataloging-in-Publication Data
A catalog record has been requested for this book

ISBN: 978-0-367-51160-9 (hbk)
ISBN: 978-0-367-51172-2 (pbk)
ISBN: 978-1-003-08151-7 (ebk)

Typeset in Times New Roman
by codeMantra

Contents

Illustrations

Figures

Tables

Acknowledgements

We wrote this book as part of the international research project GLOBUS: Reconsidering European Contributions to Global Justice, which was funded by the EU research programme Horizon 2020 from 2016 to 2020, project ID 693609. We want to emphasise that EU institutions or actors did not interfere with our research and that the aim of the project was to critically engage with EU global policies, which we hope this book shows we did.

Many thanks go to Helene Sjursen as the scientific coordinator of the project for all her encouragement and constructive criticism. Likewise, the conversations with other project members helped us a lot to formulate our approach and finetune our argument. Thus, our thanks also go to Elin Lerum Boasson, Marco Borraccetti, Michela Ceccorulli, Mai'a Davis Cross, Erik Oddvar Eriksen, Enrico Fassi, Giorgio Grappi, Melanie Hoewer, Sonia Lucarelli, Kjartan Mikalsen, Pundy Pillay, Johanne Døhlie Saltnes, Ivor Sarakinsky, Annika Styczynski, Nikola Tomic, Ben Tonra, Alexa Zellentin, Antonio Zotti, and everyone else. Furthermore, a project such as this one could not operate without excellent administrative staff. Geir Kværk has been the perfect coordinator, and Ragnhild Grønning and Trym Nohr Fjørtoft have done a great job in pushing us to make our research accessible. In Tübingen, Petra Nickisch dealt with all the administrative issues that an EU project entails with incredible patience and humour. Our student assistants Lea Augenstein, Nicolas Gäckle, Marie Groenewald, and Lukas Grundsfeld helped us a lot with our research, from compiling data to formatting the final manuscript: thank you!!

We also benefitted a great deal from workshops in Berlin, Oslo, and Tübingen; from the project reviews; and from the input we received from discussions at various conferences and the International Relations Colloquium in Tübingen. Our thanks go to Lisbeth Aggestam, Hanna Ahlström, Karin Bäckstrand, Katja Biedenkopf, Barry Buzan, Oriol Costa, Balkan Devlen, Susanne Dröge, Claire Dupont, Robert Falkner, Christine Hackenesch, Andreas Hasenclever, Benno Keppner, Stephan Keukeleire, Markus Lederer, Eva Lövbrand, Nick Mabey, Matt McDonald, Katharina Meissner, Lukas Meyer, Stephen Minas, Sebastian Oberthür, Sadhbh O'Neill, Chukwumerije Okereke, Gabi Schlag, Simon Schunz, Yvon Slingenberg,

Dennis Tänzler, Arlene Tickner, Ivo Wallimann-Helmer, Lutz Weischer, Richard Whitman, and many others whose questions pushed us to revise our argument. We are also grateful to our informants for finding time in their busy schedules to talk to us and share their experiences and knowledge, and to the students who attended Franz and Thomas's seminar on Global Climate Justice at the University of Tübingen in winter 2019/20 for their challenging questions and fruitful discussions.

As always, while many people have had their input in writing this manuscript, any errors and mistakes are of course completely our own!

Tübingen and Oslo, September 2020
Solveig Aamodt, Bettina Ahrens, Thomas Diez, Franz von Lucke

Abbreviations

AOSIS	Alliance of Small Island States
APP	Ability to Pay Principle
AWG-KP	Ad Hoc Working Group on Further Commitments for Annex 1 Parties Under the Kyoto Protocol
AWG-LCA	Ad Hoc Working Group on Long Term Cooperative Action
BASIC	countries Brazil, South Africa, India, and China
BPP	Beneficiary Pays Principle
CBDR	Common But Differentiated Responsibilities
CEEC	Central and East European countries
COP	Conference of the Parties to the UNFCCC
CVF	Climate Vulnerable Forum
DG	Directorate-General of the EU Commission
EC	European Community
ECJ	European Court of Justice
EEAS	European External Action Service
EP	European Parliament
EPSC	European Political Strategy Centre
ETS	Emissions Trading System
EU	European Union
FFF	Fridays for Future
GCCA	Global Climate Change Alliance
GCCA+	Global Climate Change Alliance Plus
GDN	Green Diplomacy Network
GHG	Greenhouse Gas
G77	Group of 77
HAC	High Ambition Coalition
ICC	International Criminal Court
INDC	Intended Nationally Determined Contribution
IPCC	Intergovernmental Panel on Climate Change
IR	International Relations
LDC	Least Developed Countries
NAMA	Nationally Appropriate Mitigation Action
NAP	National Adaptation Plans

NDC	Nationally Determined Contribution
NGO	Non-Governmental Organisation
PA	Paris Agreement
PPP	Polluter Pays Principle
RC	Respective Capabilities
R2P	Responsibility to Protect
SIDS	Small Island Developing States
TNC	Transnational Companies
TFEU	Treaty on the Functioning of the European Union
UK	United Kingdom
UN	United Nations
UNFCCC	United Nations Framework Convention on Climate Change
US	United States

1 Introduction

Normative power Europe, the liberal order, and global climate justice

1.1 The EU, normative power, and climate change

In few global policy arenas has the European Union (EU), together with its member states, been more active than in the field of climate change. While the United States (US) was the initial driving force for the adoption of international environmental policies in the late 1970s and 1980s, its interest in leading a global climate regime waned towards the end of the 1980s, and EU – until 1993, European Community (EC) – leadership became increasingly important. Indeed, following the Maastricht Treaty, climate change turned into one of the core international policy fields that the newly formed EU got involved in. Although it was certainly not the only important actor in getting the Kyoto Protocol agreed on and ratified, the decisive role of the EU in this process is undisputed (Wurzel and Connelly 2011b; Wurzel, Connelly, and Liefferink 2017; Adelle, Biedenkopf, and Torney 2018). Its internal Emissions Trading System (ETS), in effect since 2005, served as a model for emissions accounting when the 1997 Kyoto Protocol came into force in 2008. Kyoto was the first global climate agreement that included binding emission reduction targets, although one of its shortcomings was that it only compelled 38 so-called Annex I parties to such reductions. After Kyoto, the EU has made several attempts at achieving a universal regime with a more extensive reach or at least extending the duration of the Kyoto Protocol beyond the initially agreed end date of 2012.

These efforts culminated in the 2009 Copenhagen Summit, which was supposed to agree on a new climate treaty. Yet, negotiations failed and resulted in nothing more than a vague and non-binding declaration. In the years immediately following Copenhagen, the idea of a globally binding agreement was put on the back burner, and the EU stepped back from the foreground of the negotiations. When the international community eventually agreed on a new global framework in the 2015 Paris Agreement (PA), it was built on voluntary contributions and discursive pressure rather than binding rules. Yet the EU's role in achieving the PA is worth a second look. We suggest that it is perhaps better understood as the result of a learning process, which implied the acceptance of fundamental ambiguities in international society and thus led to a change of tactics, instruments, and basic understandings rather than resignation or a wholesale change of aims. The EU was important in forging the agreement in Paris, and during the negotiations it worked hard to include at least some obligatory elements. After Paris, actors within the EU started pushing for further action. In late 2019, the European Commission proposed its ambitious European Green Deal, which, if implemented, would at least in some respects restore the EU's position as a vanguard in the climate regime. However, some member state governments, including Hungary and Poland, remain sceptical and insist on a slower transformation.

How can we understand and assess the EU's global climate policy and its transformation over time? In this book, we look at this policy primarily from a *global justice* angle. We are interested in the ways in which the EU's global climate policy invokes justice concerns, implicitly or explicitly, how these concerns have shifted over time, to what effect, and how we may explain these shifts. As we will outline at greater length in *Chapter 2*, our understanding of justice focusses predominantly on procedural justice. We thus pay particular attention to the way the EU envisages the organisation of the climate regime, and the role it assigns to different actors and institutions, as well as to how it involves these actors in the negotiations towards the further development of the regime. This is different from most works on climate justice, which tend to focus on substantive justice issues, such as the distribution of responsibility for climate change mitigation (e.g. Meyer and Roser 2010; Page 2013; Caney 2014). Yet, as we argue in *Chapter 2*, the focus on substantive justice, while inspiring many concrete policy principles, all too often is mired in deliberations of an ideal world and neglects the realities of political power and negotiations in the climate regime. In addition, substantive justice claims often speak on behalf of actors who are not themselves involved in the negotiations or turn concrete individuals and groups into abstract categories. We thus think it is important to include procedural justice in the global climate justice debate. We see this book as a contribution towards this aim.

We argue that there has been a shift in EU global climate policy after the Copenhagen summit, consisting of a move away from *impartiality* as the main justice consideration. Impartiality in our framework is tied to the idea of a global, binding climate regime that takes into account intergenerational concerns and applies to state as well as non-state actors, with a neutral arbiter deciding between competing claims, and differentiations of rights and duties among regime members being the exception rather than the rule. We set impartiality against non-domination and mutual recognition (Eriksen 2016). In *non-domination*, justice is primarily guaranteed through the self-determination of states to protect the independence of their peoples. *Mutual recognition*, in contrast, demands the involvement of a variety of societal actors in decision-making as well as provisions to account for their different identities and interests.

From this angle, the PA, as well as the EU strategy in the run-up to Paris and afterwards, represents a step away from a focus on impartiality towards non-discrimination and mutual recognition. We argue that these two principles overlap in their emphasis on maintaining pluralism, but that they nonetheless stand in tension to each other. Whereas non-domination in global justice reinforces a society of states, mutual recognition works towards breaking down state boundaries and opening up space for transnational civil society engagement at various levels. The move away from impartiality thus is a move towards the enhanced recognition of the concerns of other actors, rather than assuming a congruence between the concerns of the EU and what should count as global challenges. Yet normative debate persists on whether such actors should be mainly states or, for instance, vulnerable groups.

We will expand on these justice concerns in *Chapter 2*. We propose not seeing impartiality, non-domination, and mutual recognition as opposites, but rather as three corners of a triangle that need to be kept in balance: to combat climate change fairly, we need binding rules that do not privilege some actors over others; yet we also need to take into account the diversity of histories, identities, and interests that are at stake in any global climate regime; and non-domination guards us against undue impositions of rules against countervailing claims. In that sense, as *Chapter 3* will develop, the EU's emphasis on impartiality in the 1990s and 2000s made it stand out as a progressive climate actor, but at the same time was more problematic in justice terms than is often realised, and is in part responsible for the ultimate failure to install a global climate regime. By the same token, however, we argue that the PA moved too much in the direction of non-domination to be able to achieve a just regime for all actors. While the EU has considerably improved its climate diplomacy after Copenhagen, it needs to ensure that the principle of non-domination is not becoming the overriding concern if it wants to avoid justice taking second place in the climate regime.

Despite the importance that the EU assigns to climate change and the considerable debate on climate justice in the literature, justice has so far played little to no role in analyses of EU climate policy. It is not raised much in policy assessments. Justice does not feature in the explanation of the failure of the 2009 Copenhagen summit, where the EU had strongly promoted a new binding climate treaty, or of the success of the negotiations in Paris. Instead, the existing literature on the EU and climate change has primarily taken an introspective view and focussed on the policymaking process, including the impact of interest groups (Boasson and Wettestad 2016) or the role of member states (e.g. Oberthür and Roche Kelly 2008; Oberthür and Dupont 2011). We agree, as we elaborate in *Chapter 3*, that institutional developments such as the formation of the European External Action Service (EEAS) and the competition between different EU institutions or various Directorates-General (DGs) within the Commission, have been important factors in shaping EU policy. Yet, by excluding justice concerns, such an institutionalist perspective misses an important factor in the EU's climate strategy and its influence on the global regime. Likewise, we concur with the argument that the dependence of many Central and East European countries (CEECs) on coal has shifted the priorities among member states after the 2004 enlargement, but we argue that it is problematic to phrase this only in terms of interests and not also in terms of justice claims. This book is thus a reminder to the EU climate literature to consider the importance of justice not only for global climate policies, but also for the EU and its engagement with the global climate regime.

Normative issues have of course played a role in the literature on the EU as a normative power, in which EU climate policy and in particular its role in the agreement of the Kyoto Protocol have been a major reference point (Lightfoot and Burchell 2005; Scheipers and Sicurelli 2007; van Schaik and Schunz 2012). This strand of EU Studies developed from the engagement with Ian Manners' seminal 2002 article on the EU as a 'normative power' (Manners 2002). Manners argues that the EU is different from other great powers in that its power does not rest so much on material capacities or that it is seeking material gains, but that its main asset in international politics is its ability to 'shape conceptions of "normal"' (Manners 2002, 239). The EU would thus be aiming at, and also be capable of transforming international politics from an emphasis on states and power politics to a world of regional integration, multilateral institutions, and far-reaching cooperation in the tackling of transnational problems. Among these problems was climate change, and the EU's efforts to combat climate change served as crucial support for the characterisation of the EU as a normative power. It has even led some to make the case for the EU as a 'green power' (van der Heijden 2010), although others have branded this a 'myth' (Lenschow and Sprungk 2010) and pointed to economic interests behind green policies (Falkner 2007).

Such claims to normative power rest on the EU's ability to persuade other international actors to prioritise climate change. In order to achieve this, climate change needs to be represented as a threat that overrides or is at least as important as other concerns such as economic development. Such a threat thus requires urgent countermeasures even if these compromise other policy goals. In other words, the development of an effective global climate regime presupposes the *securitisation* of climate change. Such securitisation is visible in the demands of *Fridays for Future* and other climate movements such as *Extinction Rebellion*, who have repeatedly referred to climate change as an 'emergency', 'catastrophe', or 'apocalypse'. Yet the emergency measures asked for and the demands for more urgency in implementing them also pose challenges to democracies based on deliberation and compromise (for a good discussion, see Pickering, Bäckstrand, and Schlosberg 2020). While the EU itself has securitised climate change on the global level, as *Chapter 4* shows, it has not quite done so to the extent that those marching in the streets have been doing, although the protesters have allowed EU actors to return to a more radical tone post-Paris. Crucially, the demonstrators have linked their securitisations to claims of intergenerational justice, which procedurally implies impartiality as well as mutual recognition, and in doing so have distanced themselves from the turn towards more non-domination in the PA and its focus on Nationally Determined Contributions (NDCs) to the mitigation of climate change.

Yet securitisation is not only about placing an issue on top of the policy agenda and devising measures to tackle it. In *Chapter 4*, we argue that justice claims are also linked to, and sometimes embed, securitising moves. This is because appeals to justice imply an injustice, and such instances of injustice are, in turn, linked to articulations of threat. The claims of the youth climate movements are a good illustration of this linkage. They are demands for justice for the younger and future generations, which experience injustice in the sense of current generations exploiting the resources of this planet and thus threatening the existence of humans in the future. The securitising move thus identifies an existential threat (current behaviour leading to climate change) that needs to be tackled to achieve justice. It installs a subject for whom justice is claimed through the construction of a referent object of this threat (younger and future generations). And it leads to demands for appropriate remedies to counter the threat and thus establish justice (binding mitigating measures).

We thus argue that variations in justice claims, such as those underpinning the shifts in EU climate policies, are linked to alterations of securitisation, invoking the threat of climate change in different ways and constructing different referent objects. *Chapter 4* explores the ways in which EU actors have securitised climate change, how far-reaching these securitising moves have been, how much they ran in parallel or were connected to other core EU concerns, and which of our justice principles shine through in these securitisations. Our analysis demonstrates how securitisation, on the one hand, is necessary to establish and keep climate change as a concern on the political agenda, but how, on the other hand, it may also serve as a tool for domination, imposing particular conceptions of climate change on others, and thus at the same time narrowing the debate.

1.2 Justice, the climate regime and international society

In hindsight, the normative power argument, whatever its other problems, on the one hand, was an important contribution to the development of an EU international identity as 'normative power' in the 2000s, but, on the other hand, was also a child of its times. More than the US, the EU was at the heart of the liberal moment in world order that emerged in the aftermath of the Cold War. While most analysts have emphasised the role of the US in the liberal world order, many aspects of this order, from the Responsibility to Protect (R2P) to the International Criminal Court (ICC) and the Kyoto Protocol, were seen in Washington with a great deal of suspicion and resistance, while they were at the core of EU policies. Yet rather than merely 'shaping conceptions of normal', the end of the Cold War provided a window of opportunity for the EU to promote its multilateral mode of governance, and thus the international context also shaped EU policy (Lucarelli 2018). Yet the failure of Copenhagen was an indication that *Realpolitik* had not quite disappeared from the world stage. The criticism of the Western powers' use of R2P in Libya as much as the conflicts over Crimea, Syria or Yemen, among others, brought an end to the idea of a new order, and instead reduced it to (another) liberal moment (Latham 1997). The growing rift between the EU and the US concerning the global order after the election of Donald Trump in 2016 and the continued struggles over Brexit further exemplify the limits of the liberal order and the EU's normative power.

The changing justice concerns in EU climate policy must be seen in this wider context. Impartiality, as a principle that is tied to binding rules, equal treatment of individuals, and central institution of arbitration, was very much evident in this post-Cold War liberal moment. Yet resistances to such binding rules and supranational institutions have made it clear that the 'normal' of the EU remains contested. As Richard Ashley once put it, what EU actors may regard as an urgent global concern that has to be tackled, others may see as the 'imposition' of 'international purpose' (Ashley 1989). *Chapter 5* will therefore discuss to what extent actors from the Global South have regarded the EU's policies as just or unjust. The record on this question is sobering. While EU efforts in favour of a global climate regime are welcomed by many like-minded states and especially by the most vulnerable developing countries, to others impartiality all too often looked like domination in disguise. The ETS, for instance, and the various provisions for rewarding collaborations with actors in the Global South in the Kyoto Protocol, allowed industries in Europe to continue with their emissions, and devalued emission certificates. And although the inclusion of the principle of Common But Differentiated Responsibilities (CBDR) into the United Nations Framework Convention on Climate Change (UNFCCC) and the differentiation between Annex 1 and non-Annex 1 countries in the Kyoto Protocol underlined a heightened responsibility of Northern countries, the actual emission reduction targets in the climate agreements have not sufficiently mirrored the vast disparities between the Global North and the Global South, especially if calculated in per capita emissions.

Instead, the climate agreements have mainly determined reduction targets based on the principle known as 'grandfathering', allowing industrialised states to not undermine their historically accumulated wealth. In addition, actors from the Global North, including the EU, have routinely positioned themselves against guaranteed compensations for Loss and Damage as a result of climate-induced disasters. While the climate regime itself was always to be global in nature, the negotiations over time have increasingly focussed on finding agreements with the big emitters US and later China, and gave little say to smaller countries of the Global South, let alone other societal actors. Some of these concerns may have been somewhat alleviated by the establishment of the EU's Green Diplomacy Network (GDN) in 2003 (Torney and Cross 2018), which aims at engaging in a continuous, structured dialogue with the BASIC countries (Brazil, South Africa, India and China) and other countries from the Global South, but the one-sided idea that it is the EU who shapes the normal still persists in many quarters. The postcolonial understanding of international relations informs the climate justice views and preferences of many state and non-state actors in the Global South. To them, climate change is closely connected to global historical inequality. The EU's resistance to including broader development issues in global climate governance discussions has made it difficult for the Union to generate the trust that is needed to gain a leading role in the climate negotiations.

These movements of increasing and decreasing multilateralism and responsibilities across state borders, as well as the tensions between global needs, self-determination, and the imposition of purpose are well-reflected in the conceptions of shifts between more pluralist and more solidarist versions of international society. *Chapter 6* therefore locates the EU's climate policy and the justice debates that it implies within the broader context of international society. Following Hedley Bull (1977), we see states as not simply linked by quasi-automatic, functionalist effects of their policies on other states, but by a conscious sense of sharing common, although simultaneously also contested, values and interests and thus engaging in joint institutions. The international society literature distinguishes between primary and secondary institutions (Buzan 2004). The global climate regime, with the Rio Declaration and the UNFCCC at its centre, is an example of a secondary institution with concrete, codified norms, and rules. It is situated in a set of primary institutions, understood as patterns of practice that states routinely utilise to maintain (or reinstate) order, such as diplomacy or international law.

Pluralist conceptions of international society emphasise the importance of state sovereignty and non-intervention for international order, while solidarist conceptions insist that states bear responsibilities to human beings beyond their borders. In the solidarist understanding, states thus have, in Nick Wheeler's terms, a duty to 'save strangers' (Wheeler 2000). These two conceptions of pluralism and solidarism clearly embody two of our justice conceptions: Pluralism focusses on non-domination, solidarism on impartiality. Mutual recognition, with its focus on due hearing of a multiplicity of actors would probably require a fundamental restructuring of the constituent units of international society beyond states – a move that states are unlikely to perform, as it would strip them from their (at least formal) monopoly of power in global governance. The history of the climate change regime is thus also a story of how civil society actors have been excluded and marginalised, and their access to negotiations regulated by states.

Given the EU's characterisation and ambition to act as a normative power, in particular in relation to climate change, it is no surprise that the literature on international society has seen it as an actor embodying and promoting solidarisation. Some even see 'environmental stewardship' (Falkner and

Buzan 2019) as a new institution of an increasingly solidarist international society, and the EU as a vanguard in this development, although we would caution that environmental stewardship should rather be seen as a norm than an institution in the sense of a pattern of practice. This would imply a move towards a greater importance of impartiality. As *Chapter 3* shows, the EU has indeed pushed the development of institutions that imply impartiality, for instance in its support for binding targets and the Kyoto Protocol. While this has changed after the Copenhagen summit, we show that articulations of impartiality still dominate the EU discourse; thus, the EU's strive towards more stringent rules in the post-Paris negotiations.

World history has seen different phases of more pluralist and more solidarist configurations of international society. The post-Cold War liberal moment undoubtedly was a more solidarist phase, which was followed by a movement back towards the pluralist pole of the spectrum. Pluralisation and solidarisation thus ought not to be seen as instances of teleological progress, but rather as forms of a re-balancing of fundamental normative and thus justice claims; a process that is fraught with tensions. How to guarantee the continued existence of different cultures and identities when extending forms of global governance? How to fulfil our responsibility towards others without imposing our own conceptions of life on them? If justice principles can ever only be re-balanced but not be resolved in favour of one, then this may well lead some to despair. In contrast, we argue in *Chapter 6* that these inescapable tensions may be understood as normative ambiguities that enable change. Seen from such an angle, the EU's move away from impartiality in the run-up to the PA presents itself as a recognition of the inescapable ambiguity built into international society. It thus made possible further steps towards strengthening the global climate regime, and thus towards solidarisation, while necessarily falling short of the initial aim of a forceful, binding regime. This is by no means an attempt to 'greenwash' the downsides of the PA. Instead, our emphasis on the ambiguity inherent in international society also means that we see any new re-balancing, such as in Paris, as temporary and fraught with its own shortcomings.

One of the ironies of the EU's engagement for the solidarisation of international society is that this creates tensions across different policy fields. In relation to climate change, many have pointed out that the EU's endorsement of liberal trade and the breaking down of borders in order to increase market size may be counterproductive to its climate efforts. Market-conform measures such as the ETS ultimately remain compromises that fail to deliver on the need to keep global warming below the 1.5 degrees threshold. Yet one of the EU's core solidarising strategies of enhancing the exchange of people and thus mutual recognition through increasing mobility, partnerships, and other international linkages, both within the EU and across its external borders, also runs counter to the aim of reducing emissions, even if one were to further significantly enhance trans-European rail systems.

As trade and mobility are at the heart of EU identity, it is unlikely that the EU or its member states will give up on them in order to privilege climate concerns. The resulting policies are therefore likely to balance one concern against the other: mobility must take climate change into account; climate policies must be reconcilable with market rules. Any other solution would require an even more fundamental securitisation of climate change.

1.3 EU normative power at the intersection of climate justice debates

The chapters of this book may thus be read as explorations of EU climate policy at the intersection of different political and conceptual debates in the question of global climate justice. In this sense, global climate justice is a 'nodal point' that intertwines these debates to construct a socio-political reality, while vice versa, climate justice only makes sense against the background of these debates (Laclau and Mouffe 1985; Diez 2001). We suggest that securitisation, postcolonialism, and international society are three such conceptual and political debates that come together in climate justice, and that these are crucial to make sense of EU climate policy and its shifts over time. We do not claim that these are the only discursive contexts in which EU climate policy may be understood. Visions of technological progress or narratives of economic development are undoubtedly also influencing our conceptions of global climate justice and thus of climate policy. Yet we focus on the three because they are more immediately related to climate politics. Indeed, the examples of technology and economics may be seen as further metanarratives that underpin securitisation, postcolonialism, and international society as the more directly influential discourses for climate justice.

We see the EU, which understands itself as a normative power in global climate politics, as caught in, yet also as engaging in, the normative battles that emanate from this intertwinement of discourses. Our reconstruction of the development of EU climate policies and their relation to climate justice illustrates how a specific understanding of climate justice as impartiality is at the heart of such normative power. For a long time, this understanding was not questioned within EU policymaking circles. Yet as the negotiations over a global climate agreement unfolded, it has become clear that this basis of normative power is by no means uncontested, and contestation is not only due to opposing interests but relates to different normative conceptions that are tied to different securitisations, different understandings of global power distributions, and different visions of international society. In short, as the title of this book suggests, EU normative power got caught up in normative battles, on the global level but also internally.

We trace these normative battles in core documents and speeches, supported by a series of interviews with policymakers on the EU level, in member states, and in the BASIC countries. Reading these documents from our climate justice point of view, emphasising procedural justice in the interlinkage of securitisation, postcolonialism, and international society, allows us to leave the analytical level of interest groups and institutional policy processes, and dig deeper into the normative foundations of EU climate policy that are often left untouched. In this sense, we hope that the following chapters will offer a sympathetic yet critical review of EU climate policy that opens up new possibilities to re-think the cornerstones of this policy, its contribution to the global climate regime, and its implications for global climate justice.

2 Reconfiguring the global climate justice debate

What is global climate justice? And how can we operationalise abstract philosophical concepts to study the European Union's (EU) climate policy? In this chapter, we discuss the existing literature, link it to our thematic chapters, and develop our conceptual framework of global climate justice, based on the concepts of non-domination, impartiality, and mutual recognition (Eriksen 2016). Existing debates on climate justice have mainly focussed on substantive notions of justice and abstract ideal-typical principles. We argue that, while this constitutes an important starting point, such a focus does not suffice in order to make sense of international negotiations and the climate regime. In neglecting procedural justice, the literature has not sufficiently considered the power imbalances, political constraints, and cultural differences that play a central role in climate politics. We thus make a plea for the introduction of less ideal-typical conceptions of justice that emphasise procedural aspects.

2.1 Existing approaches to climate justice: developing substantive principles

The manifold and complex intra- and intergenerational ethical implications of climate change – which Gardiner (2006) has called 'the perfect moral storm' – have fuelled a growing literature on climate justice, especially within political philosophy (e.g. Rose 1990; Shue 1993; Caney 2010; Gardiner et al. 2010; Heyward and Roser 2016). While it is beyond this chapter to discuss all aspects of these debates, we want to investigate some of their core issues to develop our own understanding of climate justice. In this chapter, we therefore take a closer look at the debates surrounding climate change as a problem of intra- and intergenerational (in)justice, and what should be the most important philosophical principles in the allocation of responsibilities for the mitigation of climate change.

2.1.1 Climate change and intra-generational, distributive, and corrective justice

A significant part of the literature focusses on climate change as a problem of *distributional* justice, which sometimes is linked to concepts such as compensatory and corrective justice (Ikeme 2003, 195, 201; Posner, Weisbach, and Sunstein 2008, 1570). From this angle, climate change brings about injustice because states and their citizens have used the common atmospheric 'sink' differently (Page 2013, 233; Blomfield 2013, 2016), while at the same time being unequally affected by the consequences of these actions (Agarwal and Narain 1991; Gardiner 2004b, 39). Thus, in order to fuel industrialisation and accumulate wealth, most developed countries have relied on fossil fuels (Okereke and Coventry 2016, 2–3) and hence have overused their 'fair share' of the atmosphere (Page 2013, 233). Because of this historical responsibility (Schüssler 2011; Meyer and Roser 2013) and their wealth, technological expertise and political capacity (Shue 1999, 534; Gardiner 2004a, 579; Caney 2005, 752; 2010, 205), these countries should be primarily held responsible for climate change and the associated abatement costs.

At the same time, developing countries have released less Greenhouse Gases (GHGs), often do not have the means to stop climate change (Gardiner 2004b, 584–85), and are more harshly affected by its effects (IPCC 2015, 15–16). This has led scholars to argue that political answers to climate change have to differentiate between 'luxury' and 'subsistence emissions' (Shue 1993). Developed countries with their luxury emissions should bear a greater part of the burden, and political solutions need to include 'antipoverty principles' and allow developing countries a 'right to develop' (Moellendorf 2015, 177–78).

In the climate regime, the principle of Common But Differentiated Responsibilities (CBDR) in the United Nations Framework Convention on Climate Change (UNFCCC) and the resulting differentiation in the obligations of states towards mitigation efforts can be seen as attempts to address these differences. As *Chapter 5* will show in more detail, distributive and corrective justice arguments still play an important role in the argumentation of developing and middle-income countries. Moreover, these questions are central when it comes to understanding instances of (indirect) domination in international society, as especially the discussion on the conception of justice as non-domination will take up later in this chapter.

2.1.2 Climate change and inter-generational justice

Due to the inertia of the climate system, climate change is also a problem of *inter-generational* (in)justice (Page 1999; Caney 2005, 749; Gardiner 2006; Moellendorf 2015). The climatic effects experienced today have been caused by past generations, while present generations create the problems of the future. Current generations will have to bear most of the immediate costs of climate mitigation (and might not be alive anymore to enjoy the results), while future generations would benefit from such actions (Gardiner 2004b, 30; Moellendorf 2015, 174). The intergenerational aspects complicate climate politics because most electoral systems and voter concerns have rather short time horizons (Gardiner 2006, 403–4). However, it is also precisely the intergenerational aspects that have inspired a range of social (youth) movements such as *Fridays for Future (FFF)* or *Extinction Rebellion* to fight for more radical and ultimately just political responses to climate change. Moreover, especially in the early days of the climate regime, intergenerational argumentations played an important role in political debates in general and in the EU's climate policy debates in particular (see *Chapter 3*).

When it comes to climate politics, intra-generational and inter-generational climate justice concerns sometimes oppose each other. Saving future generations by radically curbing emissions is likely to curtail economic development, even if one assumes that a turn to 'green economy' or 'ecological modernisation' may soften the blow somewhat (Hillebrand 2013). Unfortunately, this effect may be particularly pronounced in fragile socio-economic contexts in the Global South and hence become a problem in terms of the 'right to develop' (Ojha 2009; Moellendorf 2015, 177–78; Okereke and Coventry 2016, 3–4). This conflict of goals has contributed to creating a 'firewall' (Bodansky 2011, 4) between developed and developing countries, which has seriously hampered international negotiations. *Chapters 3* and *5* will discuss this in more detail, from the EU's perspective and from the perspective of large developing countries, respectively.

2.1.3 What should be done and who should act?

There is a consensus in the literature that while adaptation will be necessary, an effective and fair approach to climate change must primarily focus on mitigation to prevent irretrievable damage and human suffering (Ikeme 2003; Stern 2006, 1, 3; Caney 2010, 205; IPCC 2010, 2015; Zellentin 2015a). In the international negotiations, at least until the mid-2000s, the focus remained on mitigation. However, due to the growing awareness that some of the effects of climate change are already happening and because of the demands of developing countries for support in dealing with these effects, adaptation, and 'loss and damage' have increasingly become more central issues and at the same time points of contention in the negotiations, as *Chapter 5* will discuss in more detail.

Another debate has revolved around the assignment of rights and responsibilities in climate change abatement. While an individualistic perspective has its merits on a theoretical level, many scholars have taken collectives such as corporations, local communities, and especially states as reference points (Page 1999, 56, 61; Caney 2005, 757). Since such collectives exist beyond the life span of an individual, it is easier to hold them responsible for addressing climate change (Page 1999, 66; Caney 2005, 758–59; Page 2013, 232). In addition, Page (2013, 232). argues that in contrast to individuals, states possess the economic, organisational, and political means to enforce a just solution for climate change. Moreover, they are the central actors in the political negotiations on climate change. Thus, in the debate about global climate justice, a focus on states makes sense, even though there are increasing attempts to conceptualise the responsibility of corporations for climate change (Kramer and Michalowski 2012; Seck 2017).

In this book, we also focus our analysis on a collective political entity, the EU, and on states (largely China, India, Brazil, and South Africa). Yet, as we will elaborate later in this chapter, our three conceptions of climate justice emphasise the justice concerns of different subjects. While non-domination focusses on the rights of states, impartiality includes the rights and duties of individuals, whereas mutual recognition emphasises the importance of a variety of non-state actors such as indigenous people or social movements and Non-Governmental Organisations (NGOs).

2.1.4 Key principles to fairly distribute emission rights and abatement costs

Political philosophers have developed several principles for the fair distribution of the responsibility to abate climate change. The most frequently invoked one is the '*Polluter Pays Principle*' (PPP) (Caney 2005, 752; Page 2013, 237), which combines corrective and compensatory justice. In short, it holds that states that have emitted most GHGs should also clean up after themselves and hence pay for abatement measures (Caney 2005, 754; Page 2013, 237). In the climate negotiations, this principle has been central in the discussions between developed and developing countries and in the establishment of the CBDR.

Despite looking straightforward at first sight, the consistent implementation of the PPP is problematic for several reasons. As a 'backward looking approach', the PPP requires the identification of (past) polluters, and the demonstration of a direct link between the pollution and its negative effects (Page 2013, 237–38). Moreover, the question arises whether actors can be held responsible for actions the consequences of which they were not aware of at the time. Caney identifies this as the 'excusable ignorance problem': because people did not know before the 1980s – with certainty – that GHG emissions caused climate change, one may only apply the PPP for the time since the 1980s, which, in turn, would ignore large parts of historical emissions (Caney 2005, 761; 2010, 208, 210–211).

In contrast to the backward-looking PPP, the *Ability to Pay Principle* (APP) takes the current capabilities of states to contribute to mitigation and adaptation as the baseline. Yet a key objection to the APP brings back the starting point of the PPP: it ignores historical emissions and thus divides the burden between states no matter how 'dirty' their past behaviour has been (Caney 2010, 215; Page 2013, 239). On the political level, the APP has been important to legitimise calls for financial and technological assistance from wealthy developed countries. As such, it has become more relevant in the Paris Agreement (PA) negotiations and the attempts to better integrate increasingly wealthier developing countries such as China (see *Chapters 3* and *5*).

Besides the PPP and the APP, the literature has also discussed the '*Beneficiary Pays Principle*' (BPP), which does not hold those to account who polluted, but those who benefitted from the pollution. The BPP's main advantage is that it avoids the problem of having to establish a direct link between past polluters and present harmful effects of climate change (Caney 2010, 210; Page 2013, 240). In its outcome, it is therefore similar to the APP (those able to pay are likely to have benefitted from climate change), yet it does not ignore past emissions in as far as they are related to the benefits. However, the demonstration that current wealth is linked to past pollution is not that straightforward either. The BPP also entails a 'chronological unfairness' (Page 2013, 240) because it requires present beneficiaries of climate change to pay, while earlier generations that also benefited did not have to pay (Caney 2006, 473).

2.2 From non-ideal theory towards procedural justice

The debate on climate justice has been important both in clarifying the issues at stake and in moving the political agenda forward. Concepts of corrective and distributive justice, for instance, have influenced the development of the climate regime, and have shaped actor positions in the negotiations. Nonetheless, they suffer from a 'reality check' problem. On the one hand, as the discussions of the PPP, APP, and BPP have shown, concretising the more abstract principles is not an easy task, both because what makes sense in theory is not always readily translatable into political action and because while each principle makes sense on its own, their political consequences stand in tension to each other. On the other hand, the principles assume an ideal world that is a far cry from the power structures that constrain political negotiations.

Of course, most of the climate justice literature was not intended to aid an empirical analysis or give concrete policy advice (Caney 2010, 218). Instead, their main concern has been to develop sound and defendable philosophical principles that may serve as a starting point for discussion (Zellentin 2015b, 123). Yet, in recognition of the limitations of such an ideal-typical approach, to take the debate further and to better engage with practical realities, a growing literature discusses climate justice in a 'non-ideal' world (Caney 2016; Gajevic Sayegh 2016; Heyward and Roser 2016; Brandstedt 2019).

The distinction between 'ideal' and 'non-ideal theory' (Rawls 2003, 8, 216) rests on criteria such as fact-sensitivity, the level of compliance, the degree of perfection of justice principles, and its specificity (Gajevic Sayegh 2016, 3–4). In the case of climate change, the assumptions of non-ideal theories have particularly focussed on two issues: non-compliance and institutional design (Zellentin 2015b, 124–25). Concerning non-compliance, the central question is what actors should do if others do not fulfil their responsibilities in abating climate change. Should they compensate by doing more than they would otherwise be required to do, and if so, how much exactly? Or are they in this case relieved from their duties as well? (Caney 2016, 10–11). A prominent answer is Caney's 'integrationist' approach which combines taking up extra responsibilities, imposing burdens on non-compliers, and changing the incentive structures to prevent future non-compliance (Caney 2016, 13).

In relation to institutional design, the current international institutions, and also many domestic political systems seem to be poorly designed for agreeing on effective policies to curb climate change (Okereke, Bulkeley, and Schroeder 2009; Keohane and Victor 2011; Hovi, Skodvin, and Aakre 2013). The discussion about non-ideal circumstances thus leads to the question of whether it is necessary to change the rules of the game or if a focus on substantive forms of global justice suffices (Eriksen 2016, 2–4; von Lucke 2017). Following this debate, the climate justice literature stands to benefit from a greater focus on political processes and constraints, including the multitude of actors involved, the unequal distribution of various forms of political power, the role of competing and contested norms, the influence of different identities, and the acknowledgement of the diversity of political contexts in which climate policies have to be agreed upon and eventually implemented. To better understand the political negotiations and envision pathways towards progressive *and* just climate policies, we thus suggest re-adjusting the focus of research away from *what* climate justice ought to be on a substantive level towards the procedural dimension of *how* this can be achieved politically. The difficulty of the climate negotiations is not primarily finding ideal principles for distributing climate change abatement responsibilities. Instead, the challenge lies in getting political actors with different levels of political power, different identities, and different ideas about how these ideal principles should be operationalised to agree on a common, binding set of provisions (Gardiner 2006, 398). This is not to say that we discard substantive justice principles entirely. Yet we want to embed the existing substantive focus in a broader procedural framework. Thus, as the following discussion will show, some parts of our proposed framework, especially justice as impartiality, overlap with the existing theoretical principles, but cannot be reduced to them.

2.3 Three conceptions of justice: non-domination, impartiality, and mutual recognition

To broaden our understanding of climate justice, we start from a tripartite conception of global justice consisting of *non-domination*, *impartiality,* and *mutual recognition* (Eriksen 2016; Sjursen 2017). Each of these three conceptions is closely linked to specific aspects of global justice. Together they can aid us to better understand, and also give recommendations for changing the non-ideal political circumstances in which climate policies are negotiated. A key advantage of this threefold theoretical perspective lies in its avoidance of a single essentialist conception of what global justice ought to be (Sjursen 2017, 4). Instead, each of the three conceptions highlights specific aspects of justice that might be important for particular actors and issues around the world and allows for a differentiated normative discussion. Moreover, each conception comes with shortcomings and blind spots that the other two conceptions are to some extent able to address. We thus understand the three justice conceptions not as opposites, but as corners of a triangle that need to be kept in balance (see Figure 2.1 on page 30). This balance can look very differently for different aspects of the climate regime. In order to abate climate change we need shared norms and binding rules that do not privilege some actors over others; yet we also need to take into account the diversity of histories, identities, and interests that are at stake in any global climate regime; and we have to acknowledge the desire of states for self-determination and sovereignty and need to be wary of undue impositions of centralised rules against countervailing claims. In the following section, we discuss the three conceptions in more detail.

2.3.1 Global justice as non-domination

Global justice as non-domination (Eriksen 2016, 8–13) is strongly connected to neo-republican theory and its focus on freedom from domination or arbitrary interference (Pettit 1997; 2010, 140; Lovett 2009; Shapiro 2012; Bachvarova 2013). This pertains especially to forms of unauthorised or unconstitutional rule and the lack of reciprocal power (Shapiro 2012, 293–94; Sjursen 2017, 4). At the core of the conception stands the idea that states remain the key actors in the international realm and are central in causing, but also in working against, global injustices (Eriksen 2016, 11). In that sense, non-domination can be seen as a central tenet of a pluralist international society, in which states work together in institutions that primarily guarantee their own existence (Bull 1977). However, not all states are equally powerful, be it in terms of economic, military, 'soft', or 'normative' power (Nye 2004; Diez 2005; Sjursen 2006a, 2006b). In effect, this inequality can lead to 'unjust' constellations in the international system or international treaties, and to breaches of the principles of sovereignty and non-interference (Skinner 2010, 100; Sjursen 2017, 6; Eriksen 2019).

Finding ways to alleviate the influence of more powerful states, i.e. preventing them from directly or unlawfully pressuring weaker states, ignoring international agreements or dominating the outcome of the negotiations, contributes to more just solutions to global problems. Yet the conception of non-domination does not fundamentally challenge the existing pluralist order of international society (see *Chapter 6*). Instead, it aims at improving it to ensure a fair system of global governance and to prevent less powerful states from being harmed by the action or inaction of others (Eriksen 2016, 11–12). Practically, this can take different forms. Weaker states or those particularly affected by certain global problems could build coalitions or form bilateral cooperation agreements in order to withstand more powerful states and to increase their chances of getting favourable negotiation outcomes. Strengthening non-domination could also entail the support and safeguarding of international norms such as sovereignty or non-intervention, aiming to protect less powerful states from outside interference (Sjursen 2017, 7). Ultimately, however, a focus on non-domination rejects global, universalist regulations as well as supranational institutions, and instead aims to find ways to tackle global issues without relying on binding, top-down measures.

While promoting non-domination may contribute to global justice, it also has significant shortcomings (see also Markell 2008; Eriksen 2016, 12). Not all forms of domination are necessarily unjust, and leadership by certain states or coalitions is sometimes necessary for finding universally applicable international solutions to immediate collective action problems (Falkner 2007; Wurzel and Connelly 2011b; Delbeke and Vis 2015). This includes binding international treaties or supranational institutions that impose rules and behaviours onto all states. Such efforts limit the freedom of individual states and interfere with their sovereignty; hence, partly running against the idea of non-domination. Beyond these concerns, focussing only on non-domination might be insufficient for achieving global justice because it does not aim at fundamentally challenging the status quo (Cox 1981). It largely accepts the current system of states and merely aims at amending it, which neglects many injustices that are ingrained in this very system, and runs the risk of being abused by powerful states to safeguard their own sovereignty and reinforce their power position, as in the case of the United States (US)' withdrawal from multilateral institutions under the Trump Administration.

2.3.2 Global justice as impartiality

Justice as impartiality (Kane 1996; Føllesdal 2000) has its roots in Kant, natural law theorists and other proponents of universal understandings of justice and rights (Kant and Reiss 1991; Eriksen 2016, 13–18). It understands justice as a 'context transcending principle' (Eriksen 2016, 14) and emphasises the need for neutral, universalist values and institutions. These are considered just from the perspective of all involved parties and in all contexts (Sjursen 2017, 8). Moreover, justice as impartiality does not primarily aim at states as the central political subjects but focusses directly on the rights and needs of individuals, which ultimately trump the rights of states (Dworkin 2011, 333; Eriksen 2016, 14–15). Here, freedom is primarily linked to the autonomy of the individual, which goes beyond the conception of justice as non-domination (Sjursen 2017, 8). Instead of amending the existing system of states, impartiality means to actively transform this system, strengthening law-based orders to deter dominance and power inequalities, and eventually build a cosmopolitan society of individuals (Eriksen 2016, 16; Sjursen 2017, 6, 8). In the long term, disputes in the international realm are to be decided by an impartial third party, which in practice involves the strengthening of international institutions such as the United Nations (UN), the International Criminal Court (ICC), cosmopolitan law and the rights of individuals. Impartiality, as we elaborate in *Chapter 6*, thus is a core principle of a solidarist international society, in which states are not the only legitimate actors on the global level, and in which they share responsibilities for people and goods outside their own borders (Wheeler 2000). Impartiality therefore clashes with non-domination because it presupposes the possibility to interfere in other states and at the very least requires a re-definition of sovereignty as we find it, for instance, in the Responsibility to Protect (R2P) (Bellamy 2010).

On the downside, justice as impartiality cannot escape the context-bound nature of justice claims. What some would see as a neutral and universal position, will certainly be perceived as biased imposition by others (Ashley 1988; Ashley and Walker 1990), as the various disputes about the extent and applicability of human rights exemplify (Donnelly 1999, 2007). Some will therefore always perceive the strengthening of international institutions and binding global policies as a form of domination and interference in their sovereignty. In the current system of states, enhancing impartiality is thus a difficult task. Moreover, to be accepted by all or at least by most, impartial norms would always have to be phrased in fairly general terms because the more specific they get, the more resistance they will invoke (Wiener 2007). This may also include a certain ambiguity, which, however, may contribute to long-term transformation, as we discuss in *Chapter 6*. Finally, strengthening a certain set of norms that allows to interfere in the sovereignty of other states runs the risk of indeed being exploited to promote one's own, partial position in the global order, a critique often levelled, for instance, at the advancement of the liberal peace during the post-Cold War era (Chandler 2004; Richmond 2017; Chandler, Cudworth, and Hobden 2018).

2.3.3 Global justice as mutual recognition

The concept of mutual recognition (Anderson and Honneth 2005; Schmidt 2007; Eriksen 2016, 18–22) addresses the contested nature of apparently universal norms and acknowledges the Western or European bias in much of the literature underlying impartiality in particular and moral philosophy in general (Tickner 2003; Hobson 2012). Core to mutual recognition is therefore the recognition of difference (Sjursen 2017, 9). Justice is not a universal or neutral principle that applies to all and in every context in the same manner but one that has to be conceptualised as an inter-subjective category (Eriksen 2016, 20). What is just is decided not prior to the political struggles but directly in processes of deliberation among all affected parties (Young 2011; Eriksen 2016, 19). The focus is on rules of procedure that increase the legitimacy of resulting decisions by taking into account the multitude of different contexts and identities, and ensuring that all voices are heard and recognised by each other. In this respect, a particular focus of mutual recognition lies on a better representation and due hearing of formerly marginalised groups.

One problem of mutual recognition is that it is often unclear where we should draw the line between hearing all voices and actually integrating their concerns in the resulting decisions. What about positions that may themselves be normatively questionable? Furthermore, especially given the time constraints of political negotiations, recognising, and including the standpoints of all involved parties may be difficult and lead to less demanding standards that eventually only represent the lowest common denominator. This is a problem even if one excludes purely economic or privileged actors and instead focusses on formerly underrepresented groups such as indigenous people or grassroot organisations. Including all parties into negotiation processes may also underestimate the impact of existing power inequalities and the need for strong institutions to deter domination (Eriksen 2016, 22, 2019). And last but not least, the old Habermasian problem re-appears of who decides which collective identities and standpoints are legitimate and have to be included in the deliberative process. It may be important to include coal miners in the deliberations on climate change because of the possibly devastating effects that climate change mitigation efforts may have on their lives, but does the same apply to coal company executives? And how may we assess the outcomes of such deliberative processes without reference to any kind of impartial criteria? In other words, would every outcome of such negotiation processes be just, even if it fundamentally neglects the (long-term) interests of some actors or those of future generations?

2.4 Operationalising conceptions of global climate justice

Before we discuss the advantages and disadvantages of these three justice conceptions and their inter-relationship further, we explore the main characteristics of each conception of justice in the context of climate change. We discuss how they help us to see specific problems and points of contestation in the international climate negotiations, how they might facilitate more progressive and just climate policies, and what blind spots and shortcomings each conception entails concerning climate politics.

2.4.1 Non-domination and climate justice: states and coalition building

Looking at the climate regime through the lens of non-domination helps us to assess the uneven distribution of power among states and between states and other actors, the dominating effects of GHG emissions, and the role of domination in specific agreements and instruments of the regime. Such domination, for instance, persists where the rules and practices of the regime privilege large, industrialised states through their weight in negotiations and the protection of their established wealth ('grandfathering'), while climate justice principles, whether drawing on the PPP, BPP, or APP, all emphasise the special responsibility of industrialised states in the mitigation of and adaptation to climate change. Although the CBDR has tried to alleviate some of this inequality, especially concerning binding obligations to reduce GHGs, the negotiations have so far continued to privilege industrialised states (Okereke and Coventry 2016). The challenge of climate change thus raises at least two different and contradictory problematics of domination: one through necessary interference in state sovereignty, the other caused by the economic and political stratification of the international society itself. In addition, the privileging of states as legitimate actors in the international order also leads to domination in the exclusion of non-state actors or their dependence on the goodwill of states to allow their participation in the climate negotiations or assign to them roles within the regime (Araya 2015; Bäckstrand et al. 2017).

Overall, efforts to govern climate change globally, despite some solidarising tendencies and inherent ambiguities that we will discuss in *Chapter 6*, have tended to emphasise the non-domination of states in terms of safeguarding their sovereignty to the detriment of combatting structural domination and reproducing the domination of non-state actors. The UNFCCC has not specified any binding measures or sanctions to tackle climate change and has emphasised the sovereign right of all states 'to exploit their own resources pursuant to their own environmental and developmental policies' (UN 1992, 2). And even the more rigid Kyoto Protocol has calculated emission reduction targets for the participating industrialised countries applying the principle of 'grandfathering' by taking 1990 emissions as the point of reference, hence cementing the much higher per capita emissions in industrialised countries. Kyoto's 'flexible mechanisms' and the inclusion of 'carbon sinks' further favoured already powerful industrialised countries. In addition, the absence of effective sanctioning mechanisms further undermined the binding character of Kyoto so that Canada, for instance, could easily withdraw once it was clear that it would exceed its targets. Similar problems persist in the PA, which, although in principle having high ambitions and being more comprehensive than Kyoto, does not include any binding emission reduction targets (let alone sanctions) and has marginalised attempts to include binding mechanisms for the compensation of loss and damage from climate change in developing countries (Clémençon 2016, 4; Okereke and Coventry 2016, 4).

Thus, applying the principle of non-domination in a pluralist society of states has nominally safeguarded state sovereignty, but reinforced existing structures of domination to the disadvantage of small vulnerable developing countries and poor communities in coastal regions and other particularly vulnerable sites. Due to the growing global influence and GHG emissions of countries such as China, India, and Brazil, these imbalances not only concern 'North-South' relations but also the relationship between large and small developing countries i.e. 'South-South' relations (Findlay and Dimsdale 2009; Okereke and Coventry 2016, 3). While quickly becoming major perpetrators in terms of GHG output, large developing countries have been reluctant to accept their own responsibility for climate action (Christoff 2010; Paltsev et al. 2012; Hochstetler and Milkoreit 2015).

The most dramatic example of the failure to consider the interests of all states equally and avoid domination, is the situation of small island states. Despite overwhelming scientific evidence that such states are existentially threatened by the already rising sea level (Barnett and Campbell 2015; IPCC 2015, 67; Zellentin 2015a), the existing climate treaties and national pledges are not even close to preventing further sea level rise (Oels and von Lucke 2015, 64). The same goes for several geographical areas in developing countries e.g. large low-lying coastal settlements in Bangladesh (Nicholls and Cazenave 2010; IPCC 2015, 13). In addition, even if they do not disappear, most developing countries are likely to face considerable climate impacts such as droughts, floods, the spread of diseases, and food insecurity (IPCC 2015, 15–16) unless global emissions are radically curbed and financial and technological adaptation support is multiplied. Equally, future generations will face numerous climate-induced hardships if states do not quickly reduce their emissions. Not mitigating climate change in the present is a form of dominating both currently existing states and future generations. It also reproduces the uneven distribution of power in international society over time because already poor states will only get poorer when hit by climate change impacts.

A non-domination approach to climate justice that keeps the different forms of domination in balance would strive for political instruments that respect the sovereign interests of all states but also include measures that reduce the gaps in the capabilities of weaker states and other ways to alleviate stark power differences. Such an approach would largely reject legally binding international treaties that determine states contributions in a top-down manner and hence limit their freedom and sovereignty. Instead, it would welcome non-binding ad-hoc international agreements (or no agreements at all but rather declarations of intent) that set a general direction of policies but leave it to individual states to define their level of ambition in combating climate change. The hope would be that the sum of the domestically decided measures would suffice to alleviate global emission-based domination.

A non-domination regime could also include cooperation, coalitions, 'climate clubs' or bilateral agreements between some states that voluntarily decide to mitigate climate change because they are particularly affected or because they want to take on more responsibility (Narlikar and Tussie 2004; Kasa, Gullberg, and Heggelund 2008; Victor 2015). Concrete examples are the *Alliance of Small Island States* (AOSIS), the *Like-Minded Group* (Weischer, Morgan, and Patel 2012; Dröge 2016, 19), the *Climate Vulnerable Forum* (CVF), the progressive *Cartagena Group*, or the *High Ambition*

Coalition (HAC) that emerged during the negotiations of the PA. The *International Solar Alliance* initiated by India and France (International Solar Alliance 2015) or the deal on climate change and energy between the US and China (CSIS 2017) are further examples. Finally, since a non-domination approach strives to avoid too drastic regulatory interventions, it tends to rest on market mechanisms such as emissions trading schemes or other 'flexible mechanisms' as introduced by the Kyoto Protocol. These measures do not prescribe specific pathways for mitigation and give states and corporations considerable freedom in choosing the most cost-efficient ways to reduce GHGs.

Yet most of these efforts will not really be able to tackle the problem of structural domination, not to speak of the marginalisation of non-state actors. They thus demonstrate that the conception of non-domination has its limits when it comes to justice in governing a global-commons problem such as climate change. A key problem is that the freedom of one state to develop its climate policies as it sees fit, which may mean very limited mitigation, may inflict domination on other states and future generations. At the same time, states are able to dismiss calls for more radical measures as undue interference into their sovereignty. Thus, non-domination itself can turn into domination (see also Markell 2008; Eriksen 2016, 12). Coalitions of the willing may often be too weak to seriously challenge the domination of powerful state actors, and as the past has shown, market approaches to curb emissions often do not lead to decisive emission reductions (Methmann 2010; Carbonmarketwatch 2017). Ultimately, climate change may require some domination to make sure that a sufficient number of states act to alleviate global climate-related domination.

2.4.2 *Impartiality and climate justice: individuals, norms, and supranational governance*

The conception of justice as impartiality directly addresses some of the key problems of non-domination when it comes to global climate governance. Crucially, impartiality focusses on how global climate (in)justice is an issue not primarily of states but of individuals whose economic situation, political standing, and living conditions affects their climate change vulnerability (Kelly and Adger 2000; Füssel and Klein 2006). The poorest – and often politically marginalised groups such as minorities or women (Alston 2014) – will be hit first and hardest by climate change (Eakin and Luers 2006; O'Brien et al. 2007; Parry et al. 2007). This is not only true across countries, with climate change likely to have its worst impact on populations in the Global South, but also *within* more affluent countries where economically less well-off people or marginalised communities are at much higher risk to suffer from climate-induced disasters, as for instance the unequal impact of Hurricane Katrina in the US in 2005 has highlighted (Giroux 2006; Masquelier 2006; U.S. Government 2006). From the perspective of justice as impartiality, the different needs and 'human security' (Barnett, Matthew, and O'Brien 2010) of individuals cannot be ignored.

Justice as impartiality directs our attention to individuals not only as the victims, but also as the producers of climate change. It highlights how the concrete actions of individuals, their dietary choices, their means of transportation or their forms of housing play an important role in generating, but also in mitigating GHGs (Carlsson-Kanyama and Gonzalez 2009; Paterson and Stripple 2010). The accumulation of these actions leads to the vast differences in the per capita emissions of countries, which are still much higher in the Global North (Meyer and Roser 2010).

Climate justice in terms of impartiality also highlights the crucial importance of shared norms, scientific consensus (Oreskes 2004), multilateral negotiations, as well as supranational institutions in finding universally acceptable political solutions to global problems (Yamin and Depledge 2004; Ahrens 2017). To the extent that most actors within the global regime frame climate change predominantly in scientific terms, as illustrated by the central position of the Intergovernmental Panel on Climate Change (IPCC) as a body of (state-chosen) experts, they acknowledge the importance of impartiality as a core principle. However, this support quickly diminishes when it comes to actually translating these principles into concrete policies and implementing them at the national level.

So how would concrete political answers to the climate problem in line with justice as impartiality look like? In contrast to non-domination, the focus would be on partly overcoming or at least restraining the anarchy at the international level and working towards a 'solidarist' international society in which states bear responsibilities beyond their borders and individuals carry both rights and obligations on the global level (Ahrens and Diez 2015; Ahrens 2017; see also *Chapter 6*). This would primarily mean to further strengthen multilateral negotiations and work towards supranational solutions based on scientific consensus (e.g. mainly the IPCC recommendations) about adequate mitigation measures and hence determine emission reduction targets top-down rather than bottom-up, in contrast to the PA. The agreements would be as legally binding as possible, contain effective instruments to sanction non-compliance and build on the existing climate justice literature and its substantive principles to fairly cope with climate change. In that sense, of our three conceptions of justice, impartiality is closest to the discussed philosophical literature, or perhaps more accurately, most of the principles discussed in that literature presuppose a degree of impartiality, as otherwise, it is impossible to implement principles such as the PPP, BPP, or APP.

A climate regime based on justice as impartiality would also entail supporting the rights of individuals and account for the role of individuals in causing the problem. Thus, it would improve opportunities for judicial review and compensation for damages suffered (e.g. the discussions about loss and damage, funding of climate/weather insurance schemes), or expand the rights for people fleeing from climatic effects such as sea level rise (e.g. small island states inhabitants buying land elsewhere etc. (Caramel 2014). At the same time, it would increase the importance of emission allowance principles on a per capita basis (Bode 2004).

Because the measures that follow from a focus on impartiality would severely challenge the current state order, they are often legitimised through accompanying securitising moves, as we further explore in *Chapter 4*. The call for restrictions to state sovereignty or for limitations to individual freedom through the regulation of mobility or the imposition of taxes thus requires the invocation of climate change as an existential threat that warrants such measures, and their acceptance relies on the degree to which the framing as an existential threat is shared and overrides competing concerns, including economic development.

While measures linked to justice as impartiality may be able to overcome some of the problems of non-domination when it comes to facilitating progressive climate policies and global climate justice, they are no panacea. One problem is that a consensus on scientific evidence or a shared sense of the existential threat posed by climate change do not automatically translate into concrete policies. States have therefore had no problem in accepting the UNFCCC's core claim of preventing 'dangerous anthropogenic interference with the climate system' (UN 1992, 4), while the more concrete operationalisations in the Kyoto Protocol or even the PA have led to fierce opposition (Springer 2008). Because concrete policies are always more than mere one-to-one implementations of philosophical principles and instead result from power struggles and compromises, they are never entirely impartial even if they invoke impartiality (Jobst 2010). This is a particularly problematic issue in the relations between industrialised countries and the Global South, where rules seen as impartial by some are considered neo-colonial by others. We will explore this issue further in *Chapter 5*.

2.4.3 Mutual recognition and climate justice: acknowledging difference and non-state actors

A key pillar of the conception of justice as mutual recognition is its awareness of the multitude of different voices, identities, and contexts when it comes to the development and implementation of climate policies (Schroeder, Boykoff, and Spiers 2012; Schroeder and Lovell 2012). Thus, mutual recognition emphasises neither the importance of states nor of individuals. The central criterion for involving actors in deliberations is whether they are affected by climate change or by the proposed political solutions and hence have a legitimate interest in participating in the debates about adequate responses. In order to counter the danger of reproducing power inequalities in such debates, an approach informed by mutual recognition is particularly concerned about involving groups that so far have been marginalised in the climate negotiations. Mutual recognition thus to some extent follows a Foucauldian ethics that centres on the oppressed, weak, or subjugated (Pickett 1996, 460–61). In our case these are mainly non-state actors such as grassroots activist groups, indigenous peoples, and other minorities (Tsosie 2007; Schroeder 2010).

Looking at the international climate negotiations, mutual recognition urges us to acknowledge different perspectives and circumstances if we want to come to a broadly accepted political solution. It thus tries to address the conflict of interest between distributional justice in the present, especially in terms of having the freedom to expand the economy to overcome poverty, and intergenerational justice, which eventually entails reducing current emissions to a degree that could curtail economic growth (Shue 1993). From the perspective of mutual recognition, we cannot ignore either of these concerns because both represent legitimate interests of present or future actors that are seriously affected by the resulting decisions. This being said, it is of course difficult to afford due hearing to people who do not yet exist. In practical negotiations, they need to be represented by proxies who will always extrapolate their own interests into the future. Even if we assume that social movements such as *FFF* act out of best intentions, their claims will always result from the bounded rationality of their own circumstances. Ultimately, therefore, claims on behalf of future generations have to resort to claims to impartiality, as they rest on presumed 'objective' knowledge that makes it possible to forecast living conditions in distant futures.

The climate policies and international agreements that have been adopted so far clearly do not reflect the interest of all affected parties, and the climate negotiations have excluded large numbers of actors with legitimate stakes in climate policies. To recap arguments we have already discussed, the current state of the regime seriously endangers the lives and wellbeing of present vulnerable groups such as the inhabitants of small island states (Zellentin 2015a) and indigenous peoples whose very survival as distinct groups may be particularly at risk because they live in fragile environments such as the arctic, rainforests, or coastal wetlands (Tsosie 2007; Zellentin 2015a). While small island states have so far lacked the power to have their demands met in the negotiations, indigenous peoples face even greater challenges in even making their demands heard as their access to the state-based negotiations is severely constrained (Schroeder 2010). Furthermore, despite a gradual rise in the number and influence of civil society organisations in the negotiations (Araya 2015; Nasiritousi, Hjerpe, and Linnér 2016; Bäckstrand et al. 2017), they still have limited procedural rights and possibilities to shape the outcomes of these negotiations.

Moreover, many states fall short of including relevant civil society actors or hearing their suggestions in domestic climate policy processes such as the development of Nationally Appropriate Mitigation Actions (NAMAs) and Nationally Determined Contributions (NDCs) (van Tilburg, Rawlins, and Bhasin 2016; Upadhyaya et al. 2018). Since 2017, social (youth) movements such as *Extinction Rebellion* and *FFF* have demonstrated growing dissatisfaction with current policies and explicitly pointed to intergenerational injustice. While their impact on concrete policies has been limited, they at least have started to influence agenda-setting and thus have demonstrated the widespread transnational support for climate mitigation measures and the possibilities to exert pressure from outside the formal negotiations, even though these cannot replace a more systematic involvement of civil society groups to satisfy the demands of mutual recognition.

Approaching climate justice from the perspective of mutual recognition would therefore primarily take aim at fundamentally reforming the set-up of the negotiations and existing institutions at the international and domestic level. It would strive for establishing fair and inclusive negotiation and voting procedures, involving non-state actors both in the negotiations and in the administration of the regime. Domestic climate policy processes, too, would need to accommodate for different cultural and societal viewpoints. A specific focus would lie on acknowledging voices from the Global South in the negotiations, and especially from marginalised groups such as indigenous peoples. Bringing in their perspectives may not only improve the local implementation of climate change abatement projects and help to prevent new injustices (Finley-Brook and Thomas 2011; Aguilar-Støen and Hirsch 2015), but also lead to new approaches to climate change mitigation. For instance, studies find that community-owned forests – a form of land ownership that is particularly common in indigenous communities – are more sustainable than those that are privately owned (Agrawal and Ostrom 2001; Chhatre and Agrawal 2009). Beyond bringing in new ideas, such a reform of procedures could counter scepticism or a fear of neo-colonial tendencies and foster a sense of ownership among the participants that eventually could facilitate the implementation in diverse contexts.

Moreover, solutions in line with mutual recognition would have to recognise the different ideas, capabilities, and circumstances of all participants. While the CBDR or the NDCs in the PA are starting points in this regard because they do recognise difference and favour bottom-up approaches, they do not go far enough especially when it comes to giving different actors more influence in the negotiations. They mainly apply to states, are voluntary and hence overlap with principles linked to non-domination. Yet at least they force dominant players to recognise alternative framings of climate change as well as alternative policy approaches and may serve as platforms to involve a variety of actors to take part in the deliberation of national targets. In the following chapters, we thus understand both CBDR and NDCs as partly strengthening mutual recognition, depending on the context in which they are used. Ultimately, we distinguish between two forms of mutual recognition: a 'weak' one that transforms processes of deliberation in order to better recognise difference, incorporate the voices of the most affected, and foster a dialogue about appropriate solutions, but does so largely through statist frameworks; and a 'strong' one that also incorporates a broader spectrum of actors beyond the state.

Yet, just as the other two conceptions, mutual recognition is not without flaws when it comes to its implementation in the climate field. A key concern is that the negotiations, even with improved procedures, will never come close to an ideal-typical Habermasian non-authoritative deliberation. They will always be influenced by underlying relations of power and hence the results of these deliberations may always be questioned regarding their representativeness (Schroeder, Boykoff, and Spiers 2012; Augenstein 2019). As discussed earlier, it is often difficult to decide which actors have to be included. Without any reference to an at least approximately universal, impartial criterion, it will be difficult to assess whether the outcomes of more inclusive deliberations are an improvement when it comes to abating climate change fairly. Merely trying to increase the inclusiveness of the negotiations and respecting different perspectives, just as in the case of non-domination, may lead to an acceptance of the lowest common denominator and ultimately very unambitious policies. The strong focus on including all positions and a journalistic ethos for 'balanced' reporting in US climate debates, for instance, has unduly strengthened minority climate sceptical positions and eventually prevented progressive climate policies (McCright and Dunlap 2011; Eshelman 2014; Pazzanese 2016).

2.5 The triangle of climate justice and the need for balancing

Table 2.1 summarises the core characteristics of our three conceptions of justice when applied to climate change. At this point it is important to clarify that although we draw on broader philosophical and theoretical traditions with regards to non-domination, impartiality, and mutual recognition, our operationalisation of the three conceptions in the climate field is not an exact reproduction of these traditions. Instead, it is an approximation of some of their core traits in order to make them empirically observable, which undoubtedly overemphasises some aspects and downplays others. We are aware of the fact that it constitutes a theoretically less rigorous approach than usually taken in the literature on climate justice, yet we also think that such an approach is necessary and helpful to better understand the functional and ethical implications of different aspects of the climate regime in a thoroughly non-ideal political environment.

This tripartite conceptualisation of justice offers at least three distinct advantages for the analysis of the international climate regime, its key milestones, and specific policies.

First, it overcomes the narrow focus on substantive and ideal-typical climate justice principles that is present in most of the existing philosophical literature, thereby improving the possibilities for empirical application. Our approach does so by strengthening the procedural aspects of justice and putting more emphasis on the political constraints in international society, yet without giving up substantive aspects and connections to the existing literature entirely.

Table 2.1 Three Conceptions of Global Climate Justice

Dimension of Justice	*Climate Change Application*
Non-domination	– Prevent or limit the domination by powerful states – Strengthen the capabilities of (weaker) states to alleviate power inequalities e.g. through coalition building, bilateral agreements – Focus on (voluntary) market mechanisms – Avoid supranational agreements, focus on non-binding/voluntary solutions
Impartiality	– Universal solutions/principles to tackle climate change – Taking into account the rights of present and future generations – Strengthen multilateralism (supranational, binding, ambitious regimes) – central role of scientific expertise – Recognise rights and duties of individuals
Mutual recognition	– Due hearing and recognition of a range of different actors – (Fundamentally) reform procedures towards fairness, inclusiveness, and transparency – Improve process of deliberation (support smaller delegations, voting procedures) – Weak variant: focus is still on states and their different capabilities, involve non-state actors through states – Strong variant: transform negotiations to include a variety of actors beyond states

Second, we also contribute to the existing Political Science and International Relations (IR) literature on the international climate negotiations in general and the role of Europe in particular. Instead of mainly focussing on types of leadership, effectiveness, norms versus interests or on broad normative questions whether certain policies are to be considered good or bad, the tripartite perspective allows for more fine-grained and measured conclusions about EU policies. We highlight the diversity of pathways towards global climate justice and an effective regime and how actors might favour very different approaches, which can hinder or facilitate the negotiations. This approach also allows us to specifically assess the similarities and differences in terms of the organisation of key milestones of the regime.

Third, it helps us to identify and understand crucial shifts in EU policy towards the global climate regime over time. As all chapters will show in one way or another, the climate strategies of the EU and large developing countries, but also the organisation of the international climate regime itself, have undergone decisive transformations between the establishment of the regime in the early 1990s and the adoption of the PA in 2015. Our approach does not only help us to highlight these changes but also to draw conclusions of what they mean for the regime politically and ethically.

As *Figure 2.1* exemplifies, we ultimately understand the three conceptions of climate justice as the poles of a triangle. Moving from on pole to another strengthens specific justice concerns but at the same time weakens other. Thus, we understand the space between the poles as a continuum rather than as clear-cut thresholds. We also argue that it is not an equilateral triangle but that the conceptions of non-domination and mutual recognition are closer together and both further away from impartiality. Strengthening mutual recognition by better integrating different opinions and non-state actors does not necessarily lead to more domination, at least as long as it does not directly infringe on the sovereignty of states. Likewise, ensuring that no state gets unfairly dominated, does not have to limit mutual recognition, as long as non-state actors still have a say within these states. In contrast, more impartiality, hence a stronger, more binding supranational regime and a stronger role of universal norms always weakens both non-domination and mutual recognition because it overrides differences and limits the influence of states.

We think that in order to effectively and fairly abate climate change by reaching the 1.5 degree target, preventing human suffering and involving affected actors in climate negotiations, which we have taken as our normative yardstick, we need a balance between all three conceptions and the associated policy or regime components. Strengthening either of the three conceptions in isolation and too strongly, will lead to contestation and to normatively problematic extremes. In the case of non-domination, this could mean allowing countries to do nothing to abate climate change; looking at impartiality it could entail imperialistic tendencies of the supranational regime and the disregard for local concerns; and when it comes to mutual recognition, too inclusive procedures could stall the negotiations or facilitate the adoption of the lowest common denominator. Yet the appropriate

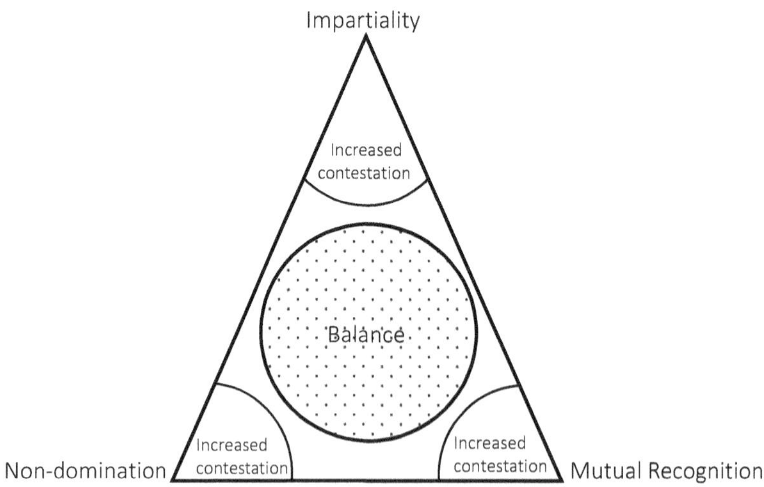

Figure 2.1 The Climate Justice Triangle.

balance does not necessarily lie in the middle of the triangle. Instead, depending on the specific problem, policy, or time period, the politically and ethically appropriate balance may be found closer to one pole than to the others. In addition, there is no 'optimal' solution, and so each balance will rightly be challenged by alternative conceptions. Climate change abatement, in this sense, remains a continuous struggle. Our plea is to recognise this struggle between different justice conceptions and the need to engage in constructive balancing – something, as *Chapters 3* and *5* will show, the EU has not always done, to say the least.

Before we move on to this analysis of EU policies over time, we should clarify that we use the tripartite framework in two ways in this book. First, we use it as an analytical tool to make sense of different aspects of the climate regime and individual policies. Thus, we draw on the three conceptions to distinguish between different pathways to tackle climate change and to understand how different actors have put emphasis on those different aspects and how this has led to points of contention or alliances in the international negotiations. Second, we draw on the tripartite framework to discuss normative questions and assess how to eventually reach the 1.5 degree target, prevent human suffering, and move the climate regime towards more inclusion.

3 Charting the development of EU involvement in the global climate regime

3.1 EU climate policy and justice conceptions

The ethical complexity of climate change, which we have summarised in *Chapter 2*, has made it difficult for political actors to come up with effective and just policies to combat climate change. The European Union (EU) has at least consistently tried to pursue ambitious climate policies and thus to navigate the 'moral climate storm'. Over the years, it has exerted a considerable degree of leadership in terms of its own climate policies and has been crucial in moving international climate negotiations forward (Falkner 2007; Schreurs and Tiberghien 2007; Oberthür and Roche Kelly 2008; Wurzel and Connelly 2011b). It was instrumental in establishing the United Nations Framework Convention on Climate Change (UNFCCC), the Kyoto Protocol, and the Paris Agreement (PA), and has also promoted provisions that acknowledge the different responsibilities of developing and developed countries, despite its increasing scepticism towards the 'firewall' created by the principle of Common But Differentiated Responsibilities (CBDR; see *Chapter 2*) (Harris 2002a; Dröge and Geden 2015, 2016).

This engagement in favour of effective and just global climate policies fits the idea of the EU as a 'normative power', outlined in *Chapter 1*, which does not primarily look after its own interests but strives for progressive and multilateral solutions to the climate problem. Time and again, EU actors have thus presented the Union as a global climate forerunner. Former Commission President Jean-Claude Juncker for instance argued:

> We Europeans are the world leaders on climate action. It was Europe that brokered the first-ever legally binding, global climate deal. It was Europe that built the coalition of ambition that made agreement in Paris possible.
>
> (Juncker 2016)

The aim of this chapter is to critically assess the EU's contribution to global climate justice. It does so by discussing two interrelated questions. First, based on the theoretical approach to global justice developed in *Chapter 2*, we explore which conceptions of justice have played a role in the EU's climate strategy, and how and why they have changed over time. Second, we analyse to what extent the EU has been able to influence the international negotiations towards a just approach towards climate change. While it is difficult to univocally single out the EU's influence on the international negotiations or prove direct causal links, we argue that we can make a plausible case for the EU's contribution by carefully tracing back key developments and policy measures in the climate regime and comparing them to the EU's positions.

Our investigation starts with the international negotiations on climate change in the 1980s but specifically focusses on the crucial period between the third Conference of Parties to the UNFCCC (COP) in Kyoto in 1997 and COP-21 in Paris in 2015. It is primarily based on a qualitative content analysis of 56 key documents issued by the European Commission, the Council of the European Union, the European Council, and the European Parliament (EP) between 1988 and 2019 (see Annex). In our analysis, we applied codes that we derived from the three overarching justice conceptions developed in *Chapter 2* as specified in *Table 3.1*. We used our coding to determine the frequency with which the three conceptions appear in EU documents. At the same time, we accounted for the importance of specific conceptions within the setup of each document (where they appear in the document, with what level of intensity they are proposed by the authors), how the standing of these documents was in relation to the eventual policies, and how the secondary literature has assessed related arguments.

In addition to this content analysis, to further substantiate our findings, gather background information, and better understand the importance of certain policies and the underlying motivations, we conducted 15 semi-standardised expert interviews with EU officials and Non-Governmental Organisation (NGO) /think tank representatives. Most of our interviewees were people working at the Directorate-General for Climate Action (DG CLIMA), the European External Action Service (EEAS) and the European Political Strategy Centre (EPSC). The interviews took place between April 2017 and March 2018 under the condition of anonymity, and instead of recording them, we collected the key insights in semi-standardised tables.

Figure 3.1 shows the respective share of the three conceptions of justice in our document sample between 1988 and 2018. To minimise the weight of outlier years, the figures show a rolling average of five years (including the two years before and after each time point). The figure clearly shows a gradual decrease of arguments linked to impartiality (I) and – especially from the mid-2000s – a steady increase of those invoking Mutual Recognition or Non-Domination. It also demonstrates that despite the rise of the other two conceptions of justice, impartiality has remained the dominant one. In the following, we explore in detail how we can make sense of this development.

Table 3.1 Indicators for Justice Dimensions

Justice Dimension	*Operationalisation for the Analysis*
Non-domination (ND)	1 Voluntariness of measures: Flexibility, non-binding and ad-hoc measures 2 Importance of state level: Bilateral agreements, sovereignty, state coalitions, alleviation of power inequalities 3 Importance of market/economic logic: market mechanisms, cost-efficiency, competitiveness
Impartiality (I)	1 Universality and Normative Arguments: scientific consensus, broad norms and philosophical debates, recourse to moral authority 2 Supranational level: international law, multilateralism, international agreements, UNFCCC, centralised measures 3 Ambition and bindingness: ambitious and concrete targets, enforcement rules, definitive, top-down regulative, and fiscal state measures 4 Role of Individuals: judicial review, empowerment
Mutual recognition (MR)	1 Recognition of difference/importance of local level (weak MR): Listen to others, recognise differences within the EU and between Global North/South, importance of local level 2 Recognition of non-state actors: importance of civil society and indigenous actors 3 Fundamentally reform procedures: bottom-up structure, increase transparency, make all voices heart, transform voting procedures

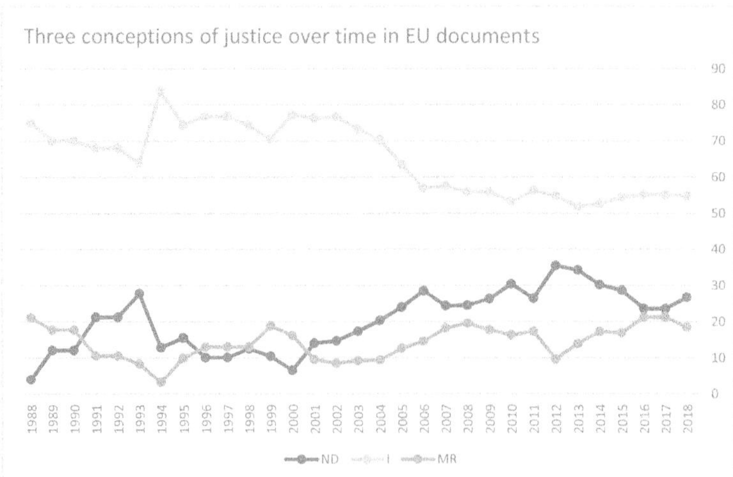

Figure 3.1 Share of the Three Conceptions of Justice Over Time (in Percentages).

3.2 The quest for an ambitious and binding climate regime

The EU has not always been a 'green normative power'. It only started to 'green' itself and to develop its actorness in environmental matters when United States (US) leadership declined and left a power vacuum in the late 1980s (Falkner 2007; Wurzel and Connelly 2011a, 3, 5). This coincided with the advent of a range of trans-border environmental problems such as ozone layer depletion and climate change (Delbeke and Vis 2015, 9). At the same time, the EU only gained competencies in environmental matters in the 1990s with the Maastricht Treaty (Delbeke and Vis 2015, 9–10). Climate change was an ideal topic to enhance the legitimacy of the newly founded EU and an opportunity to establish the Union as key player in the international sphere (Oberthür and Dupont 2011, 87). Moreover, from early on, the EU saw climate change abatement measures as an economic opportunity. It hoped to create jobs by gaining a first mover advantage in green technologies and by modernising its industries in line with the concept of 'ecological modernization' (European Commission 1994, 77; Barnes 2011, 51).

Subsequently, these drivers helped to anchor green issues at the core of EU policy and paved the way for its leadership position in the climate negotiations (Council of the European Union and European Commission 1992, 4; Falkner 2007, 509; Oberthür and Roche Kelly 2008, 36). From the beginning, the EU's approach to climate change governance was informed by a set of broad normative pillars such as multilateralism, international law, sustainable development, the belief in scientific findings and the precautionary principle (van Schaik and Schunz 2012, 173; Delbeke and Vis 2015, 8). From the perspective of our three conceptions of justice, these pillars largely correspond to an impartial understanding of global justice and emphasise the importance of supranational institutions and universal standards. They also overlap with the foundational values of the EU itself. This is a first indicator that justice concerns – even if not always explicitly cited as motivation – have always played an important role in the EU's climate strategy, although of course other more mundane political and economic interests have always been present.

Based on this normative background in line with impartiality, the EU in the 1990s emphasised the moral responsibility of present generations to quickly abate climate change in order to prevent serious harm for future people:

> We must at all costs avoid an irresponsible attitude to the management of the environment and the use of natural resources which would irreversibly jeopardise the environment that we have inherited, to the detriment of our children and our children's children.
>
> (European Commission 1991a, 2)

Moreover, it accepted the historical responsibility of developed countries (Carlarne 2010, 339; Meyer and Roser 2013) and the distributive dimension of climate change by repeatedly highlighting that 'the industrialized world was mainly responsible for CO2 emissions' (European Commission 1991a, 2) and had to act first before developing countries could be enlisted (European Commission 1991b, 12):

> There is also an increasing acceptance of a wider responsibility, as one of the foremost regional groupings in the world, to play a leading role in promoting concerted and effective action at global level, working with other industrialized countries, and assisting developing countries to overcome their special difficulties.
>
> (European Council 1990, 18)

Thus, the EU favoured an international and obligatory solution to the climate problem and acted as a 'pusher' in the international negotiations (Oberthür and Dupont 2011, 77). Consequently, one of the first climate-related Communications of the European Commission (in the following 'Commission') to the Council of the European Union ('Council' in the following) emphasised the importance of an 'international agreement for the future protection of the atmosphere' (European Commission 1988, 51) and mentioned the possibility of 'the establishment of specific targets for limiting emissions of greenhouse gases' (European Commission 1988, 51).

In the ensuing negotiations about the UNFCCC, the EU thus pressed for ambitious, concrete and legally binding emission reduction targets for the industrialised countries (Oberthür and Roche Kelly 2008, 36) on the basis of the scientific consensus (European Commission 1988, 10, 13, 38). The EU backed up this position with a pledge to stabilise its own emissions on the level of 1990 by 2000 (European Commission 1991a). It thereby demonstrated its aspiration to lead by example and to push for a political agenda (Oberthür and Dupont 2011) in line with justice as impartiality. And although the EU eventually failed to anchor binding emission reduction targets for industrialised countries in the UNFCCC, not least due to strong opposition from the US (Harris 2001, 8; Vig 2013, 91), it was instrumental in setting up the framework in 1992 and managed to place some of its core preferences on the agenda. Thus, the 1992 Rio Declaration on the UNFCCC prominently mentioned the precautionary principle (UN 1992, 9) but also the central importance of scientific findings (UN 1992, 4). Together with sustainable development (UN 1992, 10), these points are closely linked to the intergenerational aspects of climate change (van Schaik and Schunz 2012, 175) and to the universalist principles of justice as impartiality.

Even though the overall goal of the EU at that time was to make the climate regime as binding, comprehensive and ambitious as possible (European Council 1990, 19), the EU recognised the different starting positions, responsibilities and capabilities of countries from the Global South and the relevance of questions of distributional and corrective justice (European Commission 1988, 39, 41). Thus, it was in part due to the EU's influence that the final agreement on the UNFCCC included the principle of CBDR (UN 1992, 2; Carlarne 2010, 339), although its key architects were the Group of 77 (G77) and China.

As we have already discussed in *Chapter 2*, the CBDR does not easily fit our tripartite justice scheme, although it clearly does not conform to the demands of impartiality. On the one hand, it resembles a weak version of mutual recognition. It recognises the different starting positions and cultural contexts of the participants, especially of developing countries, and leaves it to the individual actor to determine their fair share in the efforts to combat climate change. In addition, the anchoring of CBDR in articles three and four of the UNFCCC constitutes a more sustained and institutionalised transformation of standard procedures than we would expect from non-domination. While the latter is primarily put into practice through ad-hoc measures such as alliances or bilateral cooperation in order to alleviate power inequalities between states, the enshrinement of CBDR in the Convention aims at permanently transforming the current order through a formal allocation of responsibility. On the other hand, this responsibility in CBDR still has states as its referent point, does not entail clear enforcement rules, and grants developing countries ample freedom. All of these three characteristics are in line with the idea of non-domination. In any case, whether CBDR implies mutual recognition or rather non-domination, the EU's support of the principle is an indication that from the start, the EU was not only concerned about impartiality in the climate regime, even if this justice conception clearly dominated its climate policy at the time.

3.2.1 Expanding leadership: the EU's role in establishing the Kyoto protocol

Despite the EU's attempts at organising the international regime in a top-down manner and making it as legally binding and ambitious as possible, this approach was not uncontested EU-internally – i.e. between the EU institutions and among different member states. This led to a considerable 'credibility gap' between the EU's external positions and internal actions (Oberthür and Dupont 2011, 76). The problem became particularly apparent during the discussions about an EU-wide carbon or energy tax, which was eventually not adopted. While the Commission favoured a tax 'based on an energy component and on a component based on carbon content' (European Commission 1991b, 9), key member states, such as the United Kingdom (UK), rejected the proposal with reference to the principle of subsidiarity (Barnes 2011, 47, 51; Oberthür and Dupont 2011, 78; Rayner and Jordan 2016, 1). The example also shows the often diverging positions of different EU institutions. Driven by the favourable public opinion on decisive climate action (Wurzel and Connelly 2011a, 9; Interview 2018f), but also due to its lack of power and hence less responsibility to actually enforce any concrete measures (Burns and Carter 2011, 59), the EP has often favoured ambitious policies that also resembled impartiality (e.g. legally binding, top-down fiscal measures). In contrast, the institutions dominated by the member states (the Council of the European Union and the European Council) have tended to prefer less prescriptive policies with more room for the sovereignty of member states (Rayner and Jordan 2016, 1), even though on the global level, they joined in the chorus of demanding an ambitious regime, supported by strong securitising moves, as *Chapter 4* will show. The Commission often was positioned somewhere in between or varied in its stance depending on its current President and composition (Barnes 2011).

Despite these internal struggles, internationally the EU continued on the route towards a supranational, binding climate regime and time and again emphasised the need for '[...] quantified objectives for significant overall reductions of Greenhouse Gas (GHG) emissions after the year 2000 below 1990 levels, within specified timeframes [...]' (Council of the European Union 1996a, 7). It hence was one of the leading actors in the negotiations towards the Kyoto Protocol that intensified in 1995 at the first COP in Berlin (Oberthür and Ott 1999; Yamin 2010). The EU emphasised the universal applicability of scientific findings and followed scientists' recommendations on necessary measures to halt global warming (Council of the European Union 1996a, 6–7, 9). It also urged to distribute the required emission reduction targets globally through a robust, top-down supranational regime (Interview 2018i). Thus, in 1996, the EU was among the first to clearly acknowledge the scientific findings of the Intergovernmental Panel on Climate Change (IPCC) and the necessity to limit global warming below two degrees Celsius, which was an ambitious target at the time (Delbeke and Vis 2015, 7):

> Given the serious risk of such an increase and particularly the very high rate of change, the Council believes that global average temperatures should not exceed 2 degrees above pre-industrial level [...].
>
> (Council of the European Union 1996b, 7)

Once again, the EU openly based its position on intergenerational justice (European Commission 1997b, 4) by linking it to sustainable development and the precautionary principle, and continued to emphasise that the industrialised countries had to lead the way with substantial and binding measures (European Commission 1997b, 3; van Schaik and Schunz 2012, 175). The Union hence reiterated principles that strengthened justice as impartiality and saw itself as a champion of such an approach. This also included the disregard for different opinions on how to cope with climate change (Schreurs 2002, 184), thereby foreshadowing the problems of a purely 'impartial' solution and the EU's persistence on this specific strategy on the international level.

When the negotiations about the Kyoto Protocol gained momentum in the run up to COP-3 in Kyoto in 1997, the EU insisted on legally binding and ambitious reduction targets for all industrialised countries and a status of 'protocol' for the agreement as the most binding option under international law (Dröge 2016, 24). Underlining the Union's determination to lead by example, it presented the prospect of a 15 per cent emission reduction by 2010 on the condition that other industrialised countries, most importantly the US, followed suit (European Commission 1997a, 17; Oberthür and Dupont 2011, 77; Wurzel and Connelly 2011a, 4). At the same time, the EU backed up its earlier endorsement of the CBDR and an exemption of developing countries from definitive reduction targets for the time being, which was enshrined in the Kyoto Protocol in the differentiation between Annex I and Non-Annex I countries (van Schaik and Schunz 2012, 179–80). The EU thus once more combined a focus on principles in line with impartiality – especially vis-à-vis the industrialised countries – with a more differentiated, flexible, and voluntary approach in relation to the Global South, resembling mutual recognition and non-domination.

Eventually, the EU was able to enshrine several of its key demands in the Kyoto Protocol. Most importantly, binding emission reduction targets, including the highest target of −8 per cent between 1990 and 2012 for the EU-15 and the second highest of −7 per cent for the US (Oberthür and Dupont 2011, 79), and – at least on paper – compliance mechanisms for industrialised countries (Nentjes and Klaassen 2004; Tabau and Maljean-Dubois 2010). At the same time, however, the EU had to make concessions to less progressive countries, especially to the US (Gardiner 2004b, 33; Oberthür and Roche Kelly 2008, 36), such as the inclusion of the market-oriented 'flexible mechanisms' in the Kyoto Protocol, which enabled actors from industrialised countries to offset their emissions through projects in developing countries. Ironically, despite its initial opposition to these instruments, market mechanisms subsequently were to play an important role in EU policies (Schreurs 2002, 188; Wurzel and Connelly 2011a, 6).

The final text of the Protocol also specified forests as carbon sinks, which effectively allowed major industrialised countries, such as Japan, Canada, and Russia, to increase their emissions (Schreurs 2002, 189). Not least due to internal criticism of these 'loopholes' by the Parliament (European Parliament 1998a, C80/228–29), the EU was able to strengthen some of the rules governing the flexible mechanisms in order to ensure that emissions were actually reduced (Oberthür and Dupont 2011, 79) – a strategy of strengthening impartial principles 'through the backdoor', which would also play a role in future negotiations (see *Chapter 6* on the ambiguity of EU policies). Yet ultimately, these instruments pushed aspects of non-domination in a highly problematic way. Instead of preventing weaker countries from being harmed by the effects of GHG emissions or strengthening their role in the negotiations, the flexible mechanisms primarily favoured the economic interests of powerful countries by weakening the binding nature of the Kyoto Protocol and the incentives to take mitigating measures at home (Barrera and Schwarze 2004; Methmann 2010). Consequently, the flexible mechanisms considerably limited the impartial character of the Protocol and also undermined its ambition to effectively address intergenerational and distributive justice (Gardiner 2004b, 25, 33).

3.2.2 *The limits of impartiality: the crisis and resurrection of Kyoto*

The negotiations about the Kyoto Protocol exemplify the ambitions of the EU to develop a regime that mainly consists of principles strengthening justice as impartiality. And despite the above-discussed concessions, the Protocol was a clear success on the part of the EU to anchor some of its preferred principles in the international regime. In the end, Kyoto emphasised scientific expertise, multilateralism, legal obligation, and largely accepted the historical responsibility of developed countries. It thereby implemented some aspects of the Polluter Pays Principle (PPP) and the Ability to Pay Principle (APP; see *Chapter 2*).

Yet Kyoto also highlights the difficulties of reaching an agreement on a global climate policy that is simultaneously effective, comprehensive, and just (Dröge 2016, 24). Especially the conflicting goals of an ambitious and legally binding agreement on the one hand (underlining principles linked to impartiality and intergenerational justice) and of exemptions for developing countries due to distributive or corrective justice considerations on the other (pointing towards mutual recognition and non-domination) gradually became problematic. While the differentiation between different groups of states derived from good intentions and legitimate practical constraints, it cemented a 'firewall' (Bodansky 2011, 4) between the Global North and South. As a consequence, these tensions soon began to undermine the effectiveness of the protocol and later the climate negotiations in general (Pauw et al. 2014, 1). They were one of the main reasons for the US refusal to ratify the protocol (Harris 2002b, 153; Kraft 2013, 112; Dröge 2016, 24) and played an important role in the eventual defection of other major emitters such as Canada, Australia, Japan, and Russia (Clémençon 2016, 5). Furthermore, the rapid growth of emerging economies such as China and India, but also Brazil, South Africa, and Indonesia increasingly blurred the distinction between developed and developing countries (Hein 2013; Dröge 2016, 15–16), making both the differentiation of responsibilities and effective mitigation even more difficult.

Paradoxically, the refusal of the US to ratify Kyoto – together with widened EU competencies on the environment – also marked the beginning of the EU's most visible and successful efforts to lead the international climate negotiations towards policies approximating justice as impartiality (Oberthür and Roche Kelly 2008, 36). Thus, due to its credible commitment to multilateralism, sustainable development, progressive climate policies and active bargaining strategies, the EU was able to salvage Kyoto (Oberthür and Roche Kelly 2008, 36; Wurzel and Connelly 2011a, 7; van Schaik and Schunz 2012, 180). In 2001, it secured the adoption of the implementation rules for Kyoto in the Marrakesh Accords, and in 2005, the protocol finally entered into force after Japan and later Russia agreed to ratify (Wurzel and Connelly 2011a, 7; UNFCCC 2013).

So, was Kyoto a success after all? Ultimately, at least the EU overachieved its original mitigation target with a reduction of 18.5 per cent between 1990 and 2012 (Delbeke and Vis 2015, 13–15; Werksman, Lefevere, and Runge-Metzger 2015, 99). However, a considerable part of this reduction was due to the de-industrialisation of former Warsaw Pact countries, which had become EU members after the 2004 'big bang' enlargement, the economic recession following the financial crisis at the end of the 2000s (Schreurs 2002, 1–2; Weidner and Mez 2008, 129; EEA 2010), and a substitution of consumer-products from Europe with products produced elsewhere, in particularly China (Karstensen, Peters, and Andrew 2018). At the same time, the emissions of large developing countries, especially China, rose ever more quickly. This has led to a steady increase of global emissions and rendered the goal to limit global warming below two degrees Celsius increasingly unrealistic (Clark 2012; Dröge 2016, 15–16). In fact, in 2007, the EU Council acknowledged that in order to reach the two-degree target, global emissions would have to be halved by 2050 – i.e. a 60–80 per cent reduction for industrialised countries (Council of the European Union 2007a) – which was a far cry from what Kyoto would have been able to achieve. Finally, as we will explore in more detail in *Chapter 6*, even the elements of Kyoto that indisputably resembled justice as impartiality, did not fundamentally challenge the status quo of a state-based global order and thus failed to contribute to the ultimate goal of an impartial understanding of justice: a solidarist international society, which places the rights of individuals at its centre (Williams 2005).

3.2.3 From Kyoto's flexible mechanisms to the EU ETS

Leaving aside the mixed impact of Kyoto on global emissions and climate justice, the inclusion of flexible mechanisms in the Kyoto Protocol laid the foundation for one of the most far-reaching EU measures to curb GHGs, the EU Emission Trading System (ETS). There is a certain irony in the fact that the EU, which originally opposed market mechanisms, was the first actor that implemented such a system. Entering into force in 2005 (Barnes 2011, 48; Meadows, Slingenberg, and Zapfel 2015, 26; Dröge 2016, 24), the ETS became the first major emissions trading system worldwide and cemented the EU's ambition to lead by example (Rayner and Jordan 2016, 7). The ETS also marked an important shift in the EU's domestic climate policies from mainly focussing on ambitious fiscal and regulatory measures (often linked to impartiality) towards a greater emphasis on recognising the differences between member states, and relying on less binding, flexible, and cost-effective instruments (European Parliament and Council of the European Union 2003, L 275/32; Oberthür and Dupont 2011, 80).

Figure 3.1 demonstrates this transformation and shows an increase of articulations linked to mutual recognition and non-domination in EU documents from the mid-2000s onwards. The rise of non-domination in our sample is partly due to an increased focus on market mechanisms, which can be attributed to Kyoto's flexible mechanisms, the EU ETS, and the necessity to 'sell' climate measures to reluctant member states. It is not non-domination in the sense that the EU opposes binding international agreements but rather the acceptance of specific instruments and a certain degree of flexibility within such binding agreements. Thus, after objections from the Council, in its first and second period of the ETS (P1: 2005–2007, P2: 2008–2012), most emission allowances were not auctioned but given gratuitously to GHG emitting industries. This comes very close to what others have called 'grandfathering', i.e. basically allowing emitters to keep emitting on a high level without much added costs (Moellendorf 2015, 177). Moreover, the ETS directive gave member states considerable freedom concerning their allocation of allowances at least in an initial three-year 'phasing in' period (European Parliament and Council of the European Union 2003, L 275/36).

Yet the ETS also contains some principles that eventually may work towards impartiality and strengthen the mutual recognition between EU member states. Not least due to pressure from the EP (Burns and Carter 2011, 61–62), an increasing percentage of the allowances are supposed to be auctioned in the periods P3–P4 (P3: 2013–2020, P4: 2021–2030). Hence, the incentive to reduce emissions is increasing. Moreover, the ETS already represents an EU-wide supranational and legally binding instrument to govern GHGs. If it was possible to stabilise the price of the allowances, this would be an effective way to lower GHGs, although in practice this has not worked because of low price tags (Carbonmarketwatch 2017). By aiming at corporations, the ETS also transcends solely state-focussed instruments, which as well approximates impartiality. Finally, the ETS uses some of its revenues for climate projects and for a modernisation fund to alleviate the 'just transition' (Interview 2018d, 2018i) from climate harming industries throughout Europe (European Commission 2016a, 4; Rayner and Jordan 2016, 11; Interview 2018h), thus respecting different circumstances and capabilities.

3.3 From Copenhagen to Paris and beyond: towards 'shared leadership'

In the years leading up to the 2009 COP-15 in Copenhagen, the EU continued to further advance the UNFCCC negotiation process and stressed the 'urgent need' (Council of the European Union 2007b, 7) for a new, global, and more encompassing climate agreement (Oberthür and Dupont 2011, 84; van Schaik and Schunz 2012, 181). Yet, by that time, the shortcomings of Kyoto had become clear and the focus was on finding a solution to the increasingly problematic distinction between developed and developing countries, and the rising share of GHGs from quickly growing emerging economies (Findlay and Dimsdale 2009, 5; Olivier et al. 2015). Thus, the EU's emphasis shifted away from measures in line with the predominately backward-looking PPP, towards the APP that considers future emission pathways and current capabilities and would require larger developing countries to mitigate as well (European Council 2007c, 13). This entailed a departure from a strict focus on principles linked to impartiality, as an inclusion of developing countries would necessitate less severe reduction regulations and alternative political measures more in line with non-domination or mutual recognition (Council of the European Union 2007b: 9).

The 2007 Bali Action Plan marked a first success in this respect, as emerging economies agreed to contribute Nationally Appropriate Mitigation Actions (NAMAs) on the condition of support from industrialised countries (UNFCCC 2009a; Hein 2013). The summit also placed adaptation measures more prominently on the UNFCCC agenda, which paved the way for the eventual integration of the principle of Loss and Damage at COP-16 in Cancún 2010 (Dröge 2016, 26–27). In general, the growing importance of adaptive and compensatory measures constitutes a move away from climate justice as impartiality. Both NAMAs and adaption measures do not impose top-down mitigation targets onto countries and are not directly linked to a universal temperature target. Instead, they are primarily defined by the different needs and preferences of countries or specific groups. In parts, such an approach corresponds to mutual recognition. However, due to the strong focus on states and the largely voluntary and non-institutionalised character, it also strengthens non-domination. And while less binding and bottom-up measures can undoubtedly contribute to global climate justice, they can also increase unfair forms of indirect domination, especially if they come with a neglect of mitigation (Moser 2012).

3.3.1 The Copenhagen summit and the defeat of the EU

Despite the gradual broadening of its position, the EU did not want to give up its main focus on mitigation and aimed at ambitious as well as binding measures for the new agreement (Delbeke and Vis 2015, 7), as the high share of argumentations linked to impartiality at the end of the 2000s demonstrates (see *Figure 3.1*). In a communication issued just before the Copenhagen summit, the Commission underlined that '[i]t is imperative to secure an ambitious outcome in Copenhagen [...]' and that '[t]o have a reasonable chance of staying below the 2°C threshold, global GHG emissions must be reduced to less than 50% of 1990 levels by 2050' (European Commission 2009b, 3). Eventually, together with a range of developing countries, the EU achieved the recognition by other key parties of the climate negotiations that climate policy should be guided by IPCC findings and should be oriented towards the two degree target (van Schaik and Schunz 2012, 181). To underline its role as leader by example and to further close the 'credibility gap' between its international proposals and domestic action (Oberthür and

Dupont 2011, 84), the EU presented a much noticed Climate and Energy Package in 2007/8 (Oberthür and Roche Kelly 2008, 41–42; Dröge 2016, 12). This contained the legally binding commitment to reduce EU emissions by 20 per cent until 2020 and a Renewable Energy Directive that aimed to increase the amount of renewables in the energy production to 20 per cent by 2020 (European Commission 2009a; Delbeke, Klaasen, and Vergote 2015, 55). The EU also pledged to commit itself to a higher reduction target and to increase its support for developing countries, if other industrialised countries followed suit (Wurzel and Connelly 2011a, 8). Moreover, the EU advocated for industrialised countries to reduce their emissions by 25–40 per cent before 2020 and 80–95 per cent by 2050 (European Council 2009b, 3; Oberthür and Dupont 2011, 84).

While this continued to strengthen an impartial conception of justice, internally, the EU began to embrace a less binding bottom-up approach and hence furthered principles in line with weak mutual recognition and non-domination. Key reasons for this change were the rounds of member state enlargement in 2004 and 2007 and the discussions about the Lisbon Treaty that led to contestations of the mantra of an 'ever closer union' (Bellamy 2013) and the continuous expansion EU competencies (Burns and Carter 2011, 70). As a reaction, the 2009 successor of the 1998 'burden sharing', now called 'effort sharing' agreement, recognised the need to balance the ambitious and legally binding aims for climate abatement against continued sustainable economic growth and the relative capabilities of the member states (European Parliament and Council of the European Union 2009, L 140/137; Rayner and Jordan 2016, 11).

Key EU policy documents thus acknowledged that '[t]he effort required of particular Member States and particular industries remains balanced and proportionate and takes their own circumstances into account' (European Commission 2008, 4) and stressed the necessity for 'an exemption [...] for sectors particularly vulnerable to competition' (European Commission 2008, 2). In general, arguments calling for a balanced approach to climate change that preserved the competitiveness of the EU economy and ensured cost-effective emission reductions increasingly gained ground (European Commission 2008, 3, 6, 10). This enhanced focus on flexibility and economic considerations was a crucial concession in order to secure the support of less wealthy member states – especially the coal-dependent Eastern European countries (Rayner and Jordan 2016, 11–12; Interview 2018h) – and to consider issues of EU internal social justice, which later were subsumed under the term of 'just transition' (Interview 2018h). It was also a reaction to the financial crisis that had begun to seriously affect several EU member states and their willingness to accept climate measures that were perceived to hurt their economies (Skovgaard 2014).

Nonetheless, at the start of the COP-15 negotiations in Copenhagen, the hopes for an ambitious and legally binding agreement were still high (Mildner and Richert 2010; Groen, Niemann, and Oberthür 2012, 178). The EU once more trusted the universal appeal of its proposed solutions to the climate problem and believed that others would be convinced by its righteousness and good example (Groen, Niemann, and Oberthür 2012, 178; Rayner and Jordan 2016, 11). However, the optimism soon faded, as the EU was unable to convince key actors of its vision and during the crucial phase of the negotiations was effectively side lined by the US and the BASIC group, consisting of Brazil, South Africa, India, and China (see also *Chapter 5*) (Curtin 2010; Oberthür and Dupont 2011, 84; Wurzel and Connelly 2011a, 8). In the end, the participants failed to agree on a new treaty. COP-15 entered the history books as one of the greatest failures of international climate diplomacy in general and of the EU's strategy and influence in particular (Groen, Niemann, and Oberthür 2012, 174). The Copenhagen Accord endorsed further action to prevent dangerous climate change. Yet it had only declaratory character, did not commit countries to a successor of the Kyoto Protocol, and lacked any specific or binding measures to abate climate change (UNFCCC 2009b). Hence, the Accord strengthened non-domination rather than impartiality, and at the same time increased the indirect domination of vulnerable countries due to continued GHG emissions.

The dynamics of the negotiations and the Copenhagen Accord itself both exemplified the problems of the EU to project its favoured approach to climate change onto the international negotiations (van Schaik and Schunz 2012, 182). The main factors that prevented a comprehensive agreement were the adherence to an explicit legal structure, which imposes binding and inflexible emission cuts onto states, as well as a lack of transparency, aside from procedural shortcomings (Monheim 2015; Walker 2018, 7, 17). Copenhagen also underlined the problems of an impartial conception of climate justice that in fact is never accepted by all as universal, especially if it is enforced by an international order and associated institutions, which many developing countries perceive as inherently unjust. This becomes particularly problematic when it is tied to a degree of self-righteousness and deafness towards diverging solutions (Donnelly 2007; Wiener 2007; Ingram 2016); hence, a lack of mutual recognition (Walker 2018) – a mode of behaviour, which was fairly apparent in the EU's approach at that time. The growing importance of the BASIC countries in the negotiations, which we will explore in more detail in *Chapter 5*, underscored the necessity to better engage with other key actors and their (often diverging) standpoints (Christoff 2010; Paltsev et al. 2012).

3.3.2 Negotiating Kyoto II and forming new alliances

The following COP-16 in 2010 and the resulting Cancun Agreements put the negotiations towards a global agreement back on track. This was to a considerable extent achieved by the Mexican Presidency's more sensitive and transparent handling of the negotiations (strengthening mutual recognition) (Randerson, Carrington, and Vaughan 2010). Notwithstanding this partial success, however, it was clear that a new global agreement would not become effective before the Kyoto Protocol expired in 2012. Consequently, the EU began to negotiate an extension of the Kyoto Protocol to a second period that would at least partly close the gap between 2012 and the commencement of a new agreement (Barnes 2011, 50). Kyoto II was finally agreed upon in the Doha Amendment at COP-18 in 2012, in which the EU reinforced its earlier target to achieve a 20 per cent reduction by 2020 (UNFCCC 2012; European Commission 2013c; Delbeke and Vis 2015, 17). While this target was still ambitious compared to other countries, the EU in fact had already reduced its emissions by 19 per cent in the first Kyoto period, so 20 per cent made little difference. On the one hand, these lowered ambitions were due to the growing realisation in the EU that 'leadership by example' and the persistence on 'targets and timetables' did not work anymore on the international level (Interview 2018c, 2018d). On the other hand, ambitious mitigation targets had become unpopular with several of the EU member states themselves (Rayner and Jordan 2016, 11–12; Interview 2018h).

Eventually, Kyoto II only covered 15 per cent of global emissions with a reduction target of 18 per cent by 2020 (Geden 2013, 1; Dröge 2016, 24). It thus was less ambitious than the original Protocol. Nevertheless, it constituted a last attempt to adopt a global, legally binding agreement, which would mainly reinforce an impartial conception of justice. Subsequently, the EU departed from this approach, at least as a standalone tactic, and increasingly also based its strategy on principles in line with mutual recognition and, if it was not avoidable, non-domination. The focus hence was more on listening to the concerns of others, fostering dialogue, and integrating different standpoints, but also on forging new strategic alliances and accepting flexible and voluntary approaches.

3.3.3 Changing strategies and strategic learning on the road towards Paris

Overcoming the distinction among the different negotiation camps and increasing its efforts in climate diplomacy was hence one of the main goals of the EU for COP-21 in Paris in 2015 (Bäckstrand and Elgström 2013; Dröge and Geden 2015, 1; Interview 2018j). Because France hosted the COP, the ambitions but also the pressure for success were high. In 2015, the EU presented as the first major economy its Intended Nationally Determined Contribution (INDC) of a 40 per cent reduction by 2030, a voluntary mechanism that had already been decided in Warsaw 2013 (Dröge 2016: 25). Yet, having realised the deficiencies of leading by example and pursuing top-down supranational solutions, the EU's own actual pledge was no longer its most important negotiation asset. In contrast to earlier negotiations, the EU was not open to increase its offer if others followed, did not insist on the legally binding term 'protocol' for the agreement, and lacked a coherent domestic package of legislation to back up its INDCs (Dröge 2016, 21–22).

Instead, the most striking feature of the EU strategy was a fundamentally changed approach to the negotiations, which we see as fostering weak mutual recognition and non-domination. This was a result of several factors originating from both the EU's internal affairs and the international sphere. Externally, the EU had to make concessions to the stronger role (and impact on the climate) of emerging economies and developing countries that would not accept a climate regime modelled after Kyoto (Qi 2011; Walsh et al. 2011; Interview 2018c, 2018h), as we discuss in *Chapter 5*. It also had to take into account the position of the US, which despite Obama's improved climate record remained opposed to far reaching, binding commitments (Goldenberg 2011). Internally, the financial crisis and the increasing contestation of previous EU climate policies, chiefly amongst new member states, also contributed to a changed EU climate strategy. It now aimed less at binding, ambitious targets and relied more on voluntary measures (Interview 2018l), which included an increasing focus on climate change as an economic opportunity (Interview 2018i) within a narrative of ecological modernisation (European Commission 1994, 77; Barnes 2011, 51). While the economic aspects had always played a role and had been further strengthened by the 2006 Stern Report, this framing became particularly strong under the Juncker Commission (2014–2019) with its emphasis on 'jobs, growth and investment' (European Commission 2014; Interview 2018k). Later, it culminated in the appointment of former deputy head of DG Trade Raffaele Mauro Petriccione as the new head of DG Clima in 2018 (Interview 2018k).

Apart from these mainly economic drivers, institutional transformations also played a role in the EU's changed approach to the climate negotiations. A central development was the growing importance of the EEAS in the EU's climate strategy (EEAS and European Commission 2013), which was a result of both the sobering diplomatic experience in Copenhagen (Jong and Schunz 2012) and the growing discussion about climate-related security implications (Solana and European Commission 2008). Closely connected, the EU upgraded the role of the Green Diplomacy Network (GDN) (EurActiv 2003), which was now led by the EEAS. The GDN was supposed to strengthen outreach activities organised by the member states' embassies around the globe in order to better understand different perspectives of third countries (Interview 2018e; Torney and Cross 2018, 45). Ultimately, these changes moved the lead on climate matters away from the Commission and towards the EEAS, especially when it came to the international negotiations (Interview 2018e), and gradually transformed climate change from an environmental towards a foreign policy issue (EEAS and European Commission 2013; Interview 2018d, 2018e).

Altogether, the EU underwent a process of 'strategic learning' (Cross 2018, 577) regarding its climate strategy and foreign policy in general (Torney and Cross 2018; Walker 2018). The changed international circumstances and the decline in global status led to a transformation of the EU's self-perception. The Union began to better understand the limits of its normative power and of its 'leadership by example' approach and acknowledged different pathways towards global climate justice (Interview 2018b). It thus moved towards 'shared' and interactive forms of leadership (Interview 2018c) and tried to negotiate the consent of others while also being open to their suggestions (Eckersley 2004). Moreover, it tried to become more realistic and pragmatic concerning its expectation for global politics, exemplified in the 2016 Global Strategy with its emphasis on 'principled pragmatism' (EEAS 2016, 8).

Thus, in the run-up to COP-21 in Paris, the EU and especially the French Presidency (Harvey 2015; Walker 2018, 9–10), engaged in a series of diplomatic efforts in order to listen to the concerns of others, and to build bridges and coalitions, especially with the developing world (Cross 2017, 3, 7, 10). As others have noticed, the focus on listening and incorporating different standpoints can be classified as a move away from justice as impartiality towards mutual recognition and a heightened awareness of the relevance of procedural justice (Walker 2018). However, the prevalent focus on states, diplomacy, coalition building, and the absence of a systematic supranational institutional framework to organise the dialogue, meant that this new strategy also contained considerable elements of non-domination.

During the COP-21 negotiations, the French Presidency thus undeniably improved the procedures and granted a greater role to non-state actors (Bäckstrand et al. 2017; Walker 2018, 9). However, the deciding players remained sovereign states (Falkner, Stephan, and Vogler 2010; Eckersley 2012), and despite increased diplomatic efforts, the negotiations took place in the old intergovernmental setting (Brugnach, Craps, and Dewulf 2017, 29). In addition, the central instrument of the PA, the Nationally Determined Contributions (NDCs), are state-based voluntary commitments that were adopted because they do not infringe much on state sovereignty (Meulemann 2015; Dröge 2016, 6).

Apart from building coalitions with key emerging economies, above all the G20 and BASIC countries, the EU strengthened its diplomatic ties with smaller developing countries by establishing partnerships with Africa, the Pacific region, and Asia (Cross 2017, 12). It reached out to key alliances in the negotiations, such as the more progressive Least Developed Countries (LDCs) and Alliance of Small Island States (AOSIS), as well as more reluctant groups such as the Group of 77 and the Like Minded Group of Developing Countries (Weischer, Morgan, and Patel 2012; Dröge 2016, 19; Interview 2018c, 2018d). In this context, the EU actively supported several countries in their efforts to develop their NDCs and National Adaptation Plans (NAPs) (Cross 2017, 17–18). A concrete example is the Global Climate Change Alliance Plus (GCCA+), through which the EU cooperates with and supports particularly vulnerable countries and also works with civil society and private actors (GCCA+ 2015b, 5; GCCA 2018a). Moreover, in its preparation for the Paris summit, the Council stressed that despite the importance of accountability, the application of the rules enshrined in a future climate agreement 'will differ according to commitment types which reflect Parties' capabilities and national circumstances' (Council of the European Union 2015, 3) and 'provide flexibility for those countries with least capabilities' (Council of the European Union 2015, 2). The same document also made clear that adaptation and 'assisting all countries, especially the poorest and particularly vulnerable ones, to achieve climate-resilient sustainable development' (Council of the European Union 2015, 2) must be a central part of the Agreement.

The EU's changed approach thus acknowledged the specific circumstances of the Global South and helped smaller delegations to cope with the often complex climate regulations and negotiations (Depledge 2016, 26; Walker 2018, 2). Yet it still focussed on states and failed to transform the UNFCCC system fundamentally or to truly integrate non-state actors, hence falling short of supporting measures in line with the strong variant of mutual recognition. The EU's ambition was to better recognise the concerns of others and at the same time to gather allies for its efforts towards a new comprehensive (but possibly also less binding) climate agreement. While this to some extent reflected a more fundamental transformation of EU values and its conception of itself as a political actor, it was also a political strategy to eventually reach a climate agreement, which would at least in parts keep an impartial vision of climate justice. The continued dominance of articulations linked to impartiality even towards the end of our period of analysis (see *Figure 3.1*), underlines these findings. We will return to this problem of advancing a strategy of impartiality in a world structured by a pluralism of sovereign states in *Chapter 6*.

3.3.4 The Paris Agreement: a success based on mutual recognition and non-domination?

By the time COP-21 in Paris kicked off, the EU, together with the French Presidency, had already built trust with many key actors. It hence laid the ground for the High Ambition Coalition (HAC) which would become a deciding factor in the negotiations (Mathiesen and Harvey 2015; Interview 2018d). The HAC bears a resemblance to both weak mutual recognition and non-domination. It is a voluntary, ad-hoc, and legally non-prescriptive alliance of states, which undoubtedly points towards non-domination. However, it only became possible in the context of diplomatic efforts that improved the due hearing of many relevant actors; hence, also strengthening mutual recognition (Schneider 2017). The EU negotiators successfully created a positive environment for the coalition and the subsequent agreement, in which nobody wanted to be left behind (Interview 2018d). This was achieved by 'working the factions' (Interview 2018c) between the key players, most notably the G77, China, and the US (Interview 2018d), and also by a flawless organisation of the negotiations and by appointing key laggards to important roles (Walker 2018, 12–15).

Eventually, on 12 December 2015, the 196 parties of the UNFCCC adopted the PA. Entering into force on 4 November 2016, the Agreement constitutes the most comprehensive climate deal so far and promises to '[hold] the increase in the global average temperature to well below 2°C above pre-industrial levels and [pursue] efforts to limit the temperature increase to 1.5°C above pre-industrial levels' (UN 2015, 3). It is also the first climate agreement that contains a definite long-term emission reduction goal (UN 2015, 4; Falkner 2016, 1115), which is derived directly from the '[...] best available science [...]' (UN 2015, 4). While the PA thus does not entirely abandon principles linked to impartiality, it lacks the centralised, top-down, and legally binding approach of the past. Instead, it mainly consists of the voluntary NDCs of the signatory states.

The NDCs are an indication of weak mutual recognition, mainly due to their bottom-up logic and their enshrinement as core mechanisms of the PA, which for instance formally integrate civil society actors in the periodic reviews of the Contributions. They thus constitute a fundamental reorganisation of key processes in the climate regime and organise emission reduction efforts according to the specific local context and capabilities of each participant. Yet these reorganisations still fail to sufficiently integrate the perspectives of indigenous peoples, other minorities or particularly vulnerable groups, and civil society (Schroeder 2010; Alston 2014; Brugnach, Craps, and Dewulf 2017). Due to these shortcomings, but particularly because of their largely voluntary character and the focus on states, the NDCs also contain several elements that point to non-domination. In the end, the NDCs are a novel operationalisation of the CBDR (see also Christoff 2016, 779) that still attempts to keep the essence of the principle. According to the PA, each participant is able to determine its contribution by 'reflecting its CBDRs and Respective Capabilities (RCs), in the light of different national circumstances' (UN 2015, 4). By not limiting this formulation to developing countries but extending it to all participants, the NDCs try to overcome the 'firewall' of the past and in terms of climate justice principles move away from the PPP towards the APP (Falkner 2016, 1115–16). While this has been a necessary step in order to get reluctant parties to sign the agreement, the reformulation of the CBDR is not unproblematic, as it reduces the weight of the historical responsibility of developed countries (Clémençon 2016, 4; Okereke and Coventry 2016, 8, 14).

At the time of ratification, the combined NDCs were far from sufficient to reach even the 2°C goal (UNFCCC 2015; Dröge 2016, 6; European Commission 2018a; Climate Action Tracker 2019). Therefore, a five-year review and an 'ambition mechanism' are supposed to ensure more aspiring targets (UN 2015, 4–5; Dröge 2016, 6, 30). These two mechanisms are legally binding and linked to universal standards, such as 'environmental integrity, transparency, accuracy, completeness, comparability and consistency' (UN 2015, 5). They therefore reinforce impartiality to some extent. However, the fact that the states were able to ascertain their initial NDCs themselves without any legally binding pressure to reach a certain ambition, significantly diminishes the impartial potential. The only pathways towards more ambition are peer as well as public pressure in the form of NGO lobbying and economic opportunities. These are supposed to 'name and shame' (Falkner 2016, 1121) participants into the right direction (Dröge 2016, 31) – a concept known from the human rights regime, the effect of which is important yet suboptimal (Risse, Ropp, and Sikkink 1999). Notably, the NDCs do not only address mitigation but also may consist mainly of adaptation measures; thus do not necessarily contribute to limit global warming. This aspect is a clear concession to climate laggards and also to many developing countries who were hoping for increased technological and financial support in the face of the already-happening adverse climatic effects (Dröge 2016, 8, 25). While broadening the scope of the climate regime and extending it with elements that support weak mutual recognition and non-domination, the NDC approach might, in the end, only lead to the smallest common denominator in terms of climate justice.

Nonetheless, a key advantage of the PA compared to Kyoto is that it was able to lift the firewall between developed and developing countries. It integrates the large emerging economies, above all China and India, and also the US, without entirely giving up the CBDR (Dröge 2016, 5, 8). Even though President Donald Trump has announced to withdraw from the PA (von Lucke 2017; Trump 2017), this has been an important step forward and exemplifies some of the advantages of overcoming a fixation on principles linked to impartiality as the only legitimate form of climate justice. Closely connected, the PA also contains elements that help to address distributional injustice and recognises the specific vulnerability of developing countries, while also ensuring continued 'sustainable development and eradication of poverty' in those countries (UN 2015, 1). Moreover, it speaks about a need for 'just transition of the workforce and the creation of decent work and quality jobs in accordance with nationally defined development priorities' (UN 2015, 2). Interestingly, the concept of just transition has been widely used in internal EU debates about how to cope with changes in the fossil fuel-based industries in some EU member states and how to 'address climate justice in a socially equitable manner' (EESC 2017; Interview 2018d, 2018h). Originally coming from trade unions, it is praised as a way to bridge climate abatement and growth. It hence resembles concepts such as ecological modernisation, industrial policy, and green economy, which have always played an important role in the EU climate discussions (Smith 2017).

In conclusion, the PA was a diplomatic success (Christoff 2016, 766; Dröge 2016, 22), which especially from the EU perspective was desperately needed (Interview 2018i). It undoubtedly renewed the belief in the UNFCCC process and reinstated the relevance of the EU. This success came with a price, though. In order to salvage a global agreement, the EU had to make considerable concessions, particularly concerning the legally binding nature of the agreement. Thus, the PA clearly departs from the top-down and binding character of Kyoto and instead mainly focusses on a voluntary, differentiated, and incremental bottom-up process; thereby strengthening both non-domination and (weak) mutual recognition. These far-reaching changes were fundamental in order to widen the scope of the agreement and EU policy experts were well aware of their crucial necessity in terms of coming to an agreement (Interview 2018e, 2018h). Yet they also make the PA vulnerable to political changes, including global external shocks such as the 2020 COVID-19 pandemic as well as the election of climate-sceptic, right-wing populist governments such as Donald Trump (Lockwood 2018).

3.3.5 EU climate policy after Paris: impartiality through the backdoor and reinforcing strong mutual recognition

Ultimately, the future of the PA crucially rests on the implementation of its measures (Interview 2018e, 2018h), continued domestic support and public pressure. After 2015, the EU has increasingly focussed on making the Paris 'rulebook' as strong as possible (Council of the European Union 2018, 7; Interview 2018c; Dröge and Rattani 2019, 5), which can be understood as an attempt to gradually bring in measures linked to impartiality through the backdoor. A 2018 document of the Council hence emphasises:

> [...] that achieving the long-term goals of the Paris Agreement is a shared commitment of all Parties and delivering on the rules and guidelines for the implementation of the Paris Agreement in a detailed, comprehensive, robust, and operational manner is a key objective of COP24.
>
> (Council of the European Union 2018, 7)

Eventually, these efforts paid off when the summit in Katowice in 2018 adopted the rulebook. Again, the EU's close engagement with other key players, especially China, but also its acceptance of differentiation when it comes to implementing the PA rules – hence weak mutual recognition – was key in securing support (Dröge and Rattani 2019, 2, 4). The rulebook itself includes some clear obligations concerning emission inventories and reporting on financial support, yet excludes anything concrete on Loss and Damage or how to reach ambitious NDCs (Dröge and Rattani 2019; UNFCCC 2019b). In the end, the Global Stocktake in 2023 will show whether the EU's efforts and the Commission's proposed European Green Deal (European Commission 2019) will further strengthen measures linked to impartiality in the future.

Besides its sustained commitment to measures in line with impartiality, the EU has also continued its emphasis on flexible and cost-effective measures (Council of the European Union 2019, 7–9), which eventually could undermine its efforts to reinstate binding measures and accountability. At the same time, EU actors have also moved towards strengthening strong mutual recognition and supported the role of non-state actors in the regime (Council of the European Union 2019, 6, 8). Thus, EU documents between 2018 and 2019 increasingly consider 'empowering citizens and non-state actors' (Council of the European Union 2019, 8) and recognise 'the importance of non-state actors and their contribution to achieve the long-term goals of the Paris Agreement' (Council of the European Union 2018, 7). This is also often linked to the UNFCCC Talanoa Dialogue, which is supposed to create a constructive atmosphere for open discussions, include all relevant stakeholders and hence embodies many characteristics of justice as mutual recognition (UNFCCC 2019a).

At COP-24 in Katowice, the Talanoa Dialogue resulted in a Talanoa Call for Action and reintroduced the HAC that already had played a key role during the Paris summit (Dröge and Rattani 2019, 4). Beyond the rise in articulations favouring measures in line with mutual recognition, the 2019 announcement of the EU to step up its spending on climate change – partly motivated by the activism of Greta Thunberg, Fridays for Future (FFF) and other social movements – underlines the hope that this will not remain mere lip-service (CNBC 2019). At the same time, the internal quarrels in 2019 over the EU's own climate targets and the implementation of the European Green Deal (Climate Home News 2019; European Commission 2019), suggest that the EU's contribution to progressive climate policies and global climate justice will remain ambiguous at best.

3.4 Conclusions: towards a diversified approach to climate justice?

In a press statement released on 8 October 2018, the IPCC issued yet another dire warning that 'limiting global warming to 1.5°C would require rapid, far-reaching and unprecedented changes in all aspects of society' (IPCC 2018). Undoubtedly, this endeavour does not only require financial and technical capabilities but also a political process that helps to create effective and just solutions to the problem. The aim of this chapter was to take a closer look at the role of the EU in building a global climate regime and to inquire whether it has lived up to its favourable image as a 'green normative power'.

We have shown that the EU's climate strategy shows a clear preference for political measures that are in line with impartiality. Particularly in the early years of the climate debates and in the negotiations leading to the UNFCCC and the Kyoto Protocol, the EU heavily relied on universalist normative argumentations. It tried to push for a legally binding global climate regime, which, based on climate science, committed industrialised countries to fixed mitigation targets in order to limit global warming below the dangerous threshold of two degrees. While the EU was fairly successful to push through this ideal in the Kyoto Protocol, the failure of COP-15 in Copenhagen in 2009 demonstrated the limits of such an approach. The summit illustrated the growing importance of large emerging economies and the need for the EU to engage with these actors. Thus, in order to salvage its own influence and also the international climate regime, the EU changed its negotiation strategy and key expectations for the regime in the run up to COP-21 in Paris in 2015. We saw this change of strategy as a form of strategic learning that not only reacted to changed external circumstances, i.e. a growing political weight of emerging economies, but also the realisation that the perspectives of others had to be listened to. At the same time, we argued that it was a result of EU-internal developments such as the resistance especially of some post-2004 member states against too ambitious, binding measures, and the strengthening and professionalisation of EU foreign policy components.

These changes do not mean that the desire for an ambitious, binding agreement disappeared entirely. However, the EU's negotiation approach increasingly integrated elements resembling justice as (weak) mutual recognition and non-domination. This mirrored changes in internal EU climate policies, which over the years had shifted from an emphasis on binding fiscal or regulatory measures towards a greater recognition of different capabilities and more voluntary market mechanisms (Oberthür and Dupont 2011, 88). Thus, after Copenhagen the EU started a range of diplomatic outreach activities, tried to better listen to the perspectives of other key actors and improved the negotiation process at the COP itself. Moreover, it abandoned its self-centred leadership by example approach and instead tried to forge progressive alliances in line with the idea of 'shared leadership'. The EU thus in parts recognised the problematic effects of unilateral leadership, i.e. exercising, often asymmetric, power over others, and instead approximated more interactive forms of leadership that try to negotiate the consent of others while also at least in parts being open to their suggestions (Eckersley 2004).

Eventually, these efforts paid off and were a crucial factor in forging the 2015 PA. While keeping a few elements that strengthen impartiality, the PA itself is mainly a hybrid between measures that fall into the mutual recognition and non-domination category. On the one hand, besides leaving a slightly bigger role for civil society, the PA's key mechanism, the NDCs, systematically transforms the climate regime, functions in a bottom up manner, and respects the different capabilities and preferences of the participants, hereby reinforcing weak mutual recognition. On the other hand, the PA's voluntary character, the focus on states, and the key role of national sovereignty when it comes to determining the individual contributions point towards non-domination. Despite a clear weakening of impartial principles, the mixture of instruments in the PA and the accompanying diplomatic strategies of the EU, especially when it comes to negotiating the Paris Rulebook, can also be read as an attempt to gradually reintroduce some impartial elements over time. Whether this strategy will pay off in the future remains to be seen.

4 Securitisation and climate justice

4.1 Securitisation and climate change

Chapter 3 has shown how the European Union (EU)'s policy towards the global climate regime has changed over time. Tracing this development, we several times have suggested that the justice claims within these policy articulations have been linked to articulations of climate change as an existential threat. This is mirrored, for instance, in the EU's 2016 Global Strategy (EUGS), which mentions the word 'climate' 24 times. When outlining the 'Priorities of Our External Action', the EUGS lists climate change as one of the core threats that 'endanger' the 'unprecedented security, democracy and prosperity' of EU citizens (EEAS 2016, 9). The strategy links climate change to increasing conflicts and 'security tensions' (EEAS 2016, 13, 27).

As previous chapters have already suggested, the security studies literature has called such a representation of an issue as an existential threat 'securitisation' (Buzan, Wæver, and Wilde 1998; Balzacq 2011). This chapter explores this link between securitisation and climate justice in more detail. Our argument is that securitisation, or rather securitising moves, are an essential aspect of climate policy in order to legitimise climate action; that this securitisation, however, takes different forms; and that these different forms imply different forms of climate justice. Thus, with the shift in its climate policies as demonstrated in *Chapter 3*, the EU also has changed its securitisation of climate change. At times, EU actors also steered clear of any explicit moves to securitise climate change, and thus 'desecuritised' the issue (Wæver 1995; Hansen 2012).

Before we unfold this argument, we need to address the question of whether it is appropriate to speak of a securitisation of climate change at all. The illustrative references from the EUGS notwithstanding, the question of whether climate change has been successfully securitised or not, has been a matter of dispute. This is because while we may find many invocations of climate change as a threat, we find few 'emergency measures' to combat that threat. This is an issue not only for the assessment of global policy outcomes, but also for the sincerity of EU securitising moves in that they may be undermined by a lack of concrete climate actions.

Following the original formulation of the so-called Copenhagen School of security studies, securitisation legitimises and therefore should at least sometimes lead to such 'emergency' or 'extraordinary' measures. The authors themselves are inconsistent in their usage when they write that securitisation 'requir[es] emergency measures and justf[ies] actions outside the normal bounds of political procedure' and then refer to 'extraordinary means' (Buzan, Wæver, and Wilde 1998, 24). This sloppiness has contributed significantly to debates over what extraordinary measures apart from the traditional security measure of war may be (Trombetta 2008).

In addition, although they explicitly reject that securitisations automatically lead to the actual implementation of emergency measures (as opposed to their legitimisation in principle), Buzan et al. (1998; Buzan and Wæver 2004) in practice take the existence of such measures as an empirical shortcut to infer successful securitisation. While this may be theoretically inconsistent, it saves them from having to go through the cumbersome process of showing how and whether individual securitising moves have indeed been successful in the sense that they have become widely accepted.

Yet do such emergency or extraordinary measures exist in combatting climate change? The answer depends of course on how one defines such measures, but on most accounts, societies and their politicians are not doing enough to tackle climate change, and thus can hardly be accused of pushing through 'emergency measures'. Even if governments arrive at policies to mitigate climate change, they normally do so after much deliberation and compromise – two central features of the kind of 'politics' that Wæver (1995) sees as overridden by securitisation. Furthermore, the extent to which the policies promoted by the EU, such as binding targets or an Emissions Trading System (ETS), really represent emergency measures is disputable. Surely, the Paris Agreement (PA) and its central mechanism of Nationally Determined Contributions (NDCs) and their periodic review is no emergency measure by any account. Indeed, the mass demonstrations organised by *Fridays for Future* (FFF) – and their respective securitisation moves by calling climate change an 'existential threat' or 'emergency' (Fridays for Future 2020) – have emerged exactly out of a frustration about the lack of decisive action against climate change, and one of the most frequently asked questions by their activists is: 'Why do you take so long?' (Prange 2019).

In their own analysis, Buzan et al. (1998) conclude that at least on the global level, climate change, and environmental issues in general, by and large have not been successfully securitised. Presumably, the authors would not come to a much different conclusion when looking at the national or regional level. Yet the argument is complicated by at least two counterarguments. One is that even in the case of military security, securitisation hardly ever leads to the *ultima ratio* of war. In fact, the whole point of deterrence is that securitisation legitimises the possibility of war but that its actual waging is averted exactly because of this possibility. So, yes, the stationing of cruise missiles during the Cold War may have been a policy that would not have been considered legitimate or even reasonable without successful securitisation, and in that sense may count at least as an extraordinary, if not as an emergency measure. Yet it did not completely override the political debate, and thus remained heavily contested to the extent that it even led to the restructuring of party systems through the rise of Green Parties (Poguntke 1987; Müller-Rommel 2002).

This directly leads to the second counterargument: governments have in fact taken many measures that would have not been considered legitimate without previous securitisation. Such measures may not immediately look like 'emergency' measures. Yet they qualify as 'actions outside the normal bounds of political procedure' (Buzan, Wæver, and Wilde 1998, 24) if those 'normal bounds' include the boundaries of the possible in political debates. Caps on emissions or compulsory building requirements may not be the desired grand policy to keep the global temperature rise below 2 or 1.5 degrees, but they are infringements on citizens' liberties that require justification. Likewise, forcing states to defend their emission plans does not as such penalise them for their lack of ambition, but it is still something that states would not be obligated to do under the conditions of a pluralist international society. One could even argue that placing climate change as an undisputed item on top of the political agenda and repeatedly discussing it in several high-level institutions such as the UN Security Council (UNSC) (UNSC 2007, 2011, 2019) has been the result of successful securitisation processes that made it clear that a denial of people-made climate change and its results is no longer an option.

In addition, analysts have pointed to the different nature of threats in different policy fields (Trombetta 2008). The threat of climate change is long-term, diffuse, and uncertain in its impact. It is a matter of probabilities. The threat of the state aggressor is immediate, visible, and clearly identifiable. Climate change is thus more often associated with risk than threat, and resilience rather than defence. This suggests that rather than of one single form of securitisation, we may speak of different kinds and degrees of securitisation and a differentiation of the types of threats between danger (concrete, immediate, visible) and risk (diffuse, uncertain, long-term) (Diez, von Lucke, and Wellmann 2016). As such, the securitisation of climate change may legitimise and sometimes even result in extraordinary measures that are not imposing extremism and violence in the same way as war does, but nonetheless represent significant changes in policy.

In terms of global climate justice, the EU climate policies are in any case first and foremost proposals and thus potential securitising moves rather than complete securitisations in the sense of having been widely accepted (Buzan, Wæver, and Wilde 1998, 25). In other words, for the purposes of our argument it is not even decisive whether climate change has been successfully securitised or not; we will instead have to ask whether EU actors have used the invocation of threats to underscore their demands, and whether these have had any justice implications. However, following from the preceding discussion, we should not judge these invocations by the extremism of the proposed measures or whether they are intended to overthrow 'normal' political procedures. We need to allow for variation not only in securitisations but already in securitising moves.

One such variation that is commonly discussed in the literature differentiates between referent objects of securitisation. A number of authors have pointed out that policy responses vary significantly depending on whether the articulated threat is supposed to endanger individual human beings, states, or planet earth as a whole, to name but three possibilities (Detraz and Betsill 2009; McDonald 2011, 2013; Diez, von Lucke, and Wellmann 2016). A second dimension refers to the articulation of a threat as concrete danger or diffuse risk (von Lucke, Diez, and Wellmann 2014; Diez, von Lucke, and Wellmann 2016). Conceptualising climate change as risk may still see climate change as an existential threat, but one that is difficult to calculate and assess, thus allowing for more options in terms of extraordinary measures than the articulation of climate change as a concrete danger to a specific group of people. This overlaps with, but is not identical to, a third dimension of time scale, which is particularly relevant to climate change. The threat may thus be immediate or medium-term, for instance in the flooding of coastal lowlands, or it may be articulated as long-term, for instance threatening the well-being of the current youth or not-even-born-yet generations.

Our argument rests on the claim that these variations in referent object, threat, and time-scale are all related to specific justice claims. Such claims build on implicit or explicit securitising moves, and the suggested remedies to injustices are legitimised by the articulation of a specific underlying form of security.

4.2 Securitisation and climate justice

Security and justice are often seen as two concepts in tension (Geyer and Guild 2016). This is because security is taken to override or marginalise justice concerns. In the terrorism debate, for instance, security has often been played off against justice in the form of individual rights. Similarly, a classic debate in conflict resolution concerns the inclusion of human rights in conflict mediation, which some see as potentially undermining more urgent physical security concerns, whereas others consider long-term security only possible if human rights are included (Babbitt 2011; Soto 2014). In Hedley Bull's classical treatment of the international society (Bull 1977), justice in terms of human rights jeopardises peace and security and thus, in Bull's view, they serve to undermine the order of the society of states.

However, in analogy to similar arguments made about the relationship between security and freedom (Neal 2011), we should not see security and justice as opposites to be weighed against each other. Instead, they are interdependent in the sense that their meaning is articulated in relation to each other, and that one presupposes the other. Justice, as freedom, is thus only thinkable on the basis of and in relation to security. And even if sometimes the two may be in tension, in other contexts, they may be intertwined or even merged to the point where justice *is* security. Ken Booth (1991), for instance, sees security as emancipation, which is at least one possible process through which justice is achieved. Likewise, the debate about human rights and conflict resolution relies on different conceptualisations of the justice-security nexus. For those who consider human rights counterproductive in conflict resolution processes, security, and justice stand in tension with each other, and at least require temporal prioritisation. Yet, to those who see human rights as an essential part of conflict resolution, security and justice are closely related to each other.

We thus cannot assume a fixed relationship between justice and security; we can only assume that they are related in one way or another. Instead, we need to trace how one is articulated through the other. Securitising moves will therefore imply specific forms of justice, and conceptions of climate justice rely on specific securitising moves that define whose justice is at stake, how deep the injustice runs, and what the time scale of addressing the injustice may be. Securitising moves as the invocation of climate change as a specific form of existential threat thus makes possible and constrains conceptions of climate justice and how it may be achieved.

Understood in such a way, securitising moves are always also justice claims. They are about recognition in the invocation of a referent object, which thus ought to be given its rightful place in the public debate. They are about injustices in the articulation of an existential threat. And they are about paths to rectify these injustices in the attempted legitimisation of extraordinary measures. Successful securitisation, thus, imposes hegemony and at least temporarily fixes the meaning not only of climate security but also of climate justice in a particular societal and historical context (Laclau and Mouffe 1985; Burgess 2011).

This conceptualisation of the relationship between justice and security differs from the debate about the extent to which securitisation may be considered just. Rita Floyd (2011, 2019b) has defended securitisation against those who, following a conceptualisation of justice and security as opposites, have seen securitisation as reproducing existing power structures and inherently conservative. Floyd instead sees securitisation as potentially emancipatory, and thus just, if it satisfies a set of three criteria: the objective existence of an existential threat, the moral legitimacy of the referent object, and an appropriate response in the extraordinary measures taken with reasonable chances of success.

Floyd's argument attempts a moral justification of securitisation based on objective criteria derived from normative theorising. We, however, treat securitising moves as putting forward particular conceptions of justice. In other words, we argue that implicitly or explicitly, actors pursuing climate policies on the global scale (which is the focus of this book) will have to rely on securitising moves that attempt to shift the debate towards measures that are not as yet common, and in doing so put forward specific climate justice claims. The pursuit of climate policies on the global level is thus not just a matter of mitigating or adapting to climate change. Climate policy, in particular with its present prominence on the political agenda, also functions as a discursive nodal point through which conceptions of security, justice, and more broadly societal order are inscribed into the broader public debate (Laclau and Mouffe 1985; Diez 2001). As such, articulations of climate policy are attempts in exercising political power (von Lucke 2020b), and the analysis of the interlinkages of climate securitisation and justice are thus central to our understanding of the role that climate policy plays in this broader debate. This includes, for instance, the architecture of North-South relations, which we will address in *Chapter 5*.

The normative assessment of different climate securitisations, following Floyd's proposal, is an additional step following this reconstruction of security- and justice-making through securitisation. This additional step is important even though the securitisation of climate change as such seems to be just in the sense that 'any organisation/state which fails to curb carbon emissions is now guilty of harmful neglect' (Floyd 2019a, 394). Yet this only assesses the first criterion of the existential threat. It still leaves questions about the referent object: Is it just, for instance, to make the state the core referent object of climate security instead of individuals? Or should present individuals be prioritised over future generations? Should the prime referent object of climate security be vulnerable communities in the Global South or a universal world society?

It is equally debatable whether the invoked extraordinary measures are 'appropriate' or not. Their appropriateness, in turn, refers back to the construction of the threat itself. For instance, is climate change unavoidable and would justice therefore, in the extreme, demand a focus on adaptation rather than mitigation? Presumably, most readers will intuitively consider adaptation without mitigation to be misguided and unjust towards future generations (Gardiner 2006). Yet comparative studies of climate securitisations have shown that such securitising moves do occur, often in relation to the state and the military capacity to defend state territory as the main referent object, for instance on the federal level in the US (Diez, von Lucke, and Wellmann 2016, 36–64). The global justice implications of such a form of securitisation are profound. They suggest non-domination as the core principle, coupled with a denial of historical responsibilities and distributive justice, especially if such justice is future-oriented and goes beyond national boundaries.

Given these considerations, we may formulate a number of expectations for the justice implications of securitisations along the three dimensions of referent object, threat, and time. Take the referent object. If securitising moves articulate the individual as threatened for instance due to food-shortages, they are likely to imply impartiality, in addition to some form of redistributive justice. Examples would be attempts to strengthen the human rights of affected individuals in climate treaties, for instance by enshrining a right to judicial review or compensation. If indigenous groups or coastal regions or a diversity of different actors with little say on the level of international society are threatened, we assume that mutual recognition becomes more central. Thus, in order to better acknowledge their specific damages due to climate change, the procedures at the Conferences of the Parties to the United Nations Framework Convention on Climate Change (COPs) would have to be transformed. And the state or state institutions as the main referent object imply non-domination.

Likewise, the more long-term the threat is constructed, the more we would expect invocations of intergenerational justice, which, as we have discussed in *Chapter 2*, rely on a degree of impartiality that is necessary to institutionalise the imaginary voice of non-present, future generations, although such a voice may also become present through a broadening of the debate towards social movements in mutual recognition. Short-term dangers and crises, by contrast, tend to trigger a resurfacing of nationalist strategies and non-domination concerns, especially if supranational structures are weakly developed, as the Corona crisis of 2020 has aptly demonstrated. Meanwhile, the justice implications of the articulations as risk or danger are less clear. Although in principle, one would assume that the more diffuse character of risk would speak in favour of mutual recognition, resilience strategies as the most common response to minimise risk or risk effects can be pursued on all levels and may, in turn, respond to individual justice as much as justice for national communities. Table 4.1 provides an overview of the different dimensions of securitisation and corresponding climate justice claims.

In the remainder of this chapter, we will draw on the empirical evidence presented in *Chapter 3* to show how EU actors up to Copenhagen were performing securitising moves that primarily invoked climate change as a universal existential threat to individuals both now and in the future, thus underpinning their strong emphasis on impartiality as the main justice principle. However, as climate justice conceptions became more varied in the post-Copenhagen period, so did climate securitisation moves, which now involved a greater variety of referent objects and threat articulations, referring especially to particularly vulnerable groups, and were at least less

Table 4.1 Typical Dimensions of Securitisation and Corresponding Climate Justice Claims

Securitising Move *Climate Justice Claim*	• *Main Referent Object*	• *Type of Threat*	• *Time Horizon*
Non-domination	States	Concrete	Short-term
Mutual-recognition	Groups	Mixed	Various
Impartiality	Individuals	Mixed	Long-term

explicit about their time horizon. We also show that, in the late 2000s and early 2010s, EU actors articulated climate change on the basis of the preceding, agenda-setting securitisation but refrained from explicit securitising moves, amounting to desecuritisation in the post-Copenhagen period in the sense that Lene Hansen (2012, 535) once described as 'strategic time-out'.

4.3 The EU, securitisation, and impartiality

The early stages of the climate change debate within EU institutions were still characterised by a significant degree of uncertainty. Even so, the documents clearly performed securitising moves. They saw climate change as dangerous, with catastrophic effects on the well-being of the population, not just in a specific region but across the globe. They identified human beings as the main source of climate change and called for urgent measures. Justice therefore always arose as truly global justice, as justice for all humans, present and future, and even the planet as a whole. This is in line with our conception of impartiality, which, as we have shown in chapter three, prevailed until the Copenhagen COP-15.

The elements of a securitising move are already visible in an early proposal for a work programme considering 'the greenhouse effect and the Community', in which the European Commission noted that 'the risks are alarmingly high and the likely direct and indirect consequences potentially disruptive' (European Commission 1988, #8.2). In line with a securitising move, the Commission invoked alarm and disruption, while at the same time speaking of 'risk' rather than 'threat' and qualifying the disruption as 'potential'. Nonetheless, the communication was written in 'a sense of urgency' generated by '[r]ecent international events' (European Commission 1988, #8.3).

Nearly two years later, in June 1990, the Dublin European Council declaration 'on the environmental imperative' (European Council 1990) used even more drastic language. It saw the 'natural environment which forms the life support system of our planet ... gravely at risk' and the 'earth's atmosphere seriously threatened' (European Council 1990, 17). The declaration articulated climate change clearly and emphatically as an existential threat, and the referent object was not merely the global population, but planet earth as a whole. The declaration emphasised that the 'quality of life – indeed, the continuation of life – could no longer be assured were recent trends to proceed unchallenged' (European Council 1990, 17). While the reference to the 'quality of life' seems to suggest human beings as a referent object, the crescendo to 'the continuation of life' is more open, and in combination with the previous sentence again suggests that climate change is threatening the entire global habitat. Likewise, the Commission in its 1991 Community Strategy to battle climate change noted that the 'greenhouse problem is global in nature' (European Commission 1991b, #4).

Consequently, the heads of government 'recognize [their] special responsibility for the environment, both to our own citizens and to the wider world'

(European Council 1990, 17). Three years later, approving the United Nations Framework Convention on Climate Change (UNFCCC), the Council adopted the phrasing in the Convention and spoke of 'dangerous anthropogenic interference with the climate system' (Council of the European Union 1993).

The European Parliament (EP) concurred. In a 1998 resolution on the 1997 Kyoto COP-3, it painted the disastrous effects of climate change in bold colours:

> 1997 was the hottest year in the century, 0.44°C warmer than the mean temperature in 1960–1990; ... crops were affected in some of the least developed countries of the world, and ... the droughts, floods, tornadoes, bushfires and the El Niño effect, experienced all over the world, gave a foretaste of the likely effects of global warming and entailed a heavy social and economic burden.
>
> (European Parliament 1998b, #G)

In the run-up to Copenhagen, similar rhetoric appeared. In a 2005 memorandum on *Winning the Battle Against Global Climate Change* (European Commission 2005), the Commission warned that '[c]limate change is a long-term threat that calls for a long-term, global response' (European Commission 2005, 2). The referent object in this memorandum remained planet earth as a whole, while at the same time the text clearly laid out the potentially disastrous consequences for human societies:

> It is projected that ecosystems will suffer from climate change, with some species and habitats disappearing. Global food production is likely to decline; infectious diseases, e.g. malaria, dengue and schistosomiasis, may spread, and water scarcity and quality is likely to be an issue in many regions. ... Weather impacts are likely to include higher maximum temperatures, more heat waves, increased summer drying with the risk of drought and fires, or, in other regions, increases in precipitation, storms and floods. Such impacts might trigger so-called secondary effects, such as regional conflicts, poverty, famine and migration.
>
> (European Commission 2005, 3)

At the same time, presumably to reinforce the sense of urgency to EU citizens, the referent object increasingly became the EU:

> We are already witnessing some impacts. The heat wave that engulfed Europe during the summer of 2003 is typical of what experts consider to be the likely effects of climate change. During that summer, more than 20,000 people in the EU died from a combination of heat stress and increased air pollution from ozone (another greenhouse gas) and particulates. Southern Europe suffered large-scale forest fires. European farmers lost over € 10 billion in income.
>
> (European Commission 2005, 3)

Interestingly, the 2003 EU Security Strategy (ESS) did not list climate change among the major threats that the Union was facing (European Council 2003). In his widely acknowledged 2008 report on the implementation of the ESS to the European Council, however, then High Representative Javier Solana claimed that the ESS had 'already identified the security implications of climate change' and saw them as having taken on additional 'urgency'. Adopting a phrase that had become popular in United States (US) think tanks debates on climate security (CNA 2007, 1, 6), Solana claimed that climate change was a 'threat multiplier':

> Natural disasters, environmental degradation and competition for resources exacerbate conflict, especially in situations of poverty and population growth, with humanitarian, health, political and security consequences, including greater migration. Climate change can also lead to disputes over trade routes, maritime zones and resources previously inaccessible.
>
> (European Union 2008, 5)

In the meantime, the European Council in its meeting conclusions had repeatedly pointed to the security implications of climate change (European Council 2005, #43, 2007a, #27, 2007b, #41). Immediately before the 2009 Copenhagen COP, the Council moved climate change up to the top of the agenda, right after 'institutional issues'. In its analysis, the 'climate is changing faster than expected and the risks this poses can already be seen. We are experiencing widespread melting of ice, rising global sea levels and increased frequency, intensity and duration of floods, droughts and heat waves' (European Council 2009a, 1).

In line with these articulations of an increasingly existential threat to a global human community, planet earth as a whole, and also the EU and its citizens, EU actors were consistently demanding urgent global action. As we have already demonstrated in *Chapter 3*, the EU in this period focussed on binding measures, from targets for industrialised countries to proposals for a carbon tax. In relation to securitisation, the repeated emphasis on the urgency of actions is noteworthy.

Thus, for instance, the 1990 European Council concluded: 'We *urge* all countries to introduce extensive energy efficiency and conservation measures and to adopt *as soon as possible* targets and strategies for limiting emissions of greenhouse gases' (European Council 1990, 19; emphasis added). In 1996, the Council 'underline[d] the need for urgent action at the widest possible level' (Council of the European Union 1996a, #3). Or in 2007:

> The challenges of climate change need to be tackled effectively and urgently. ... The European Council underlines the vital importance of achieving the strategic objective of limiting the global average temperature increase to not more than 2°C above pre-industrial levels.
>
> (European Council 2007b, #27)

Or similarly, just before the Copenhagen COP: 'all countries, including those not presently bound by the Kyoto Protocol, should take immediate action' (European Council 2009a, #6).

Yet we also need to point to references of uncertainty and ambiguity. In 1988, the Commission, for instance, noted that the 'various impacts of such climatic change and their socio-economic consequences cannot be reliably assessed in detail at present' (European Commission 1988, 6). Especially in these early years of the debate, the demand for emergency measures was not as unconditional as we would expect from a 'pure' securitisation move. Thus, for instance, the Commission's stance from the beginning was that 'a careful assessment of [the] economic viability [of measures against climate change] is required' (European Commission 1988, 7). In 1991, the Commission argued that a 'first step in controlling CO2 emissions needs to include measures which involve the lowest economic costs and which at the same time lead to benefits in other policy areas' (European Commission 1991b, 3).

However, this balancing approach received a lot of criticism from the EP as well as, perhaps more surprisingly, from the European Council, especially in response to the Commission's proposal for a Community Strategy, which to the other actors seemed too timid, in particular in its approach to the European car industry. The European Council thus was 'concerned about the possibility that the approach for a strategy proposed by the Commission may in the end not be sufficient to ensure the attainment of the average CO2-emission-value target' (Council of the European Union 1996a, 3). And the EP criticised a 1998 Commission proposal for regulating the car industry as not providing 'any satisfactory answer to a number of urgent questions which are vital for successful implementation' (European Parliament 1998b, #11).

The securitising moves in the period up to Copenhagen thus overwhelmingly invoke an existential threat, although with some uncertainties, to humankind, present and future, demanding urgent action not only from the EU but also from all global actors. In the documents there is little acknowledgement of the special responsibility of industrialised countries or a differentiated approach. The justice claims involved in these securitising moves place humanity at the centre, and thus those who are presently or will in the future be particularly affected by climate change, which is constructed as an existential threat to their very existence. Averting this threat thus became an imperative that cannot be negotiated. The European Council, for instance, explicitly declared responsibility for future generations: 'tomorrow's environment depends on how we act today' (European Council 1990, #20).

Thus, in terms of our categories of procedural justice, these securitisation moves demanded impartiality. In their perspective, it would be unjust to invoke state sovereignty and, in the spirit of non-domination, privilege one's own nation over others in the light of such a global existential threat. This becomes particularly clear in the EP's condemnation of the 'unreasonable pressures brought on the [Kyoto] Conference by the USA and by other industrialized countries' (European Parliament 1998b, 228) and the failure of the US to ratify the Kyoto Protocol (European Parliament 2002, #D).

The statements also show little sense of diverse justice claims. The variation in the effects of climate change on different groups does not imply a particular right of involvement in decision-making, but rather underscores their substantive justice claims, which may be assessed objectively and impartially by experts and integrated in the global regime, for instance through the Common But Differentiated Responsibilities (CBDR) principle, but also through special adaptation measures 'in developing countries, … focusing on countries and regions that are particularly vulnerable to the adverse impacts of climate change' (European Council 2009a, #10). Even so, particularly from 2007 after COP-13 in Bali and the adoption of the Bali Action Plan, EU actors made it clear that all countries – and especially emerging economies – need to do their bit to mitigate climate change, independent of their developmental status.

4.4 Post-Copenhagen: desecuritisation and flexibilisation

One of the most immediately obvious differences in terms of securitising moves between the pre- and the post-Copenhagen period is the almost complete absence of any invocations of threat in European Council conclusions. Whereas up to 2009, these conclusions included a great deal of securitising moves, they now still argued that global measures to combat climate change are necessary but no longer explicitly refer to any kind of threat, danger, or even risk. The European Council initially hung on to a 'global and comprehensive legal agreement' as 'the only effective way to reach the agreed objective of staying below a 2°C increase in global temperatures compared to preindustrial levels' (European Council 2010a, 11). But in the five years after Copenhagen, climate change receded into the background and became an issue predominantly within the larger context of the *2030 Climate and Energy Policy Framework* (European Council 2014a, #15, 2014b).

Similarly, Presidency conclusions of the Council of the EU adopted a nearly managerial attitude of how to best achieve the reduction of Greenhouse Gas (GHG) emissions in order to keep global warming below the two-degree threshold. They routinely reconfirm the reality of global warming, refer to the necessity to take appropriate measures, and emphasise the EU's commitment to reduce emissions by 80–95 per cent by 2050 compared to 1990 (Council of the European Union 2012, #3, 2011, #1; see European Commission 2011, 3). The closest the Council gets to a securitising move is when the conclusions raise 'concern' about the development of climate change in general or the latest reports by the Intergovernmental Panel for Climate Change (IPCC) in particular (Council of the European Union 2015, #2). Instead, issues of 'coherence, improved coordination, efficiency, effectiveness, responsiveness, balanced representation and institutional economy' become more important for instance in discussions about climate finance (Council of the European Union 2010, #6). And climate change, now a 'challenge' rather than a threat, must be addressed 'in a balanced manner' (European Parliament and Council of the European Union 2013, Rec. 4), although in fairness we need to add that the same Regulation still uses the UNFCCC phrasing of 'dangerous anthropogenic interference with the climate system' (European Parliament and Council of the European Union 2013, Rec. 3).

This near disappearance of securitising moves may in part be an effect of having achieved the initial extraordinary measure, the firm positioning of climate change on the political agenda. Thus, the necessary goals are frequently repeated, namely 'that to keep the 2°C objective within reach, global GHG emissions need to peak by 2020 at the latest and be reduced by at least 50% by 2050 compared to 1990 and continue to decline thereafter' (Council of the European Union 2012, #3).

Nonetheless, the lack of any explicit securitising moves seems initially surprising given the widely perceived failure of Copenhagen, which would have required an extra push towards moving beyond agenda-setting into 'real' action. However, COP-15 made it clear to the EU actors that the imposition of their own views may well be seen by other actors as unjust, as *Chapter 5* will explore further. In that sense, the desecuritisation may also be seen as a 'strategic time-out' (Hansen 2012, 535) in order not to offend others unnecessarily and instead focus on technical measures within the scope of the agreed. In the meantime, a basic goal of the initial securitisations, the placement of climate change on the international political agenda, had been achieved and routinised. This decline in securitising moves in the EU context after COP-15 is in line with the global climate-security debate, which after 2009 gradually began to fade in several countries (Diez, von Lucke, and Wellmann 2016). Apart from having achieved the initial goal of setting climate-change onto the political agenda, a further reason for this decline was what others have described as 'disaster or climate fatigue' (Capstick and Pidgeon 2014). After the debate had been heavily securitised in the run-up to the Copenhagen summit, yet, without reaching a substantial agreement nor with any of the drastic threats materialising, the political effect (agenda setting, generating attention, creating urgency) of constructing climate change as a threat substantially declined.

Beyond that, the climate debate also needs to be seen in the broader political and economic context. A closer look reveals the influence of the financial crisis from late 2007 onwards as well as the increasing tensions with Russia, for instance in the Georgian-Russian war of 2008, and their impact on energy supply chains. Thus, issues of financial and energy security gain in importance. Climate security is increasingly seen as intertwined with energy security and ought not to undermine economic and financial stability. In that sense, climate change is also desecuritised because it was at least partially replaced by another threat in the discourse (Hansen 2012, 541).

The Environment Council in 2011, for instance, 'recognise[d] the importance of transitioning to a low-carbon economy and society in Europe in driving sustainable growth and innovation, increasing energy security, creating jobs and enhancing competitiveness' (Council of the European Union 2011, #3). Climate protection in this view makes sense not only because it is necessary in the light of existential threats but also because it makes sense economically and energy-wise:

> We have seen an economic crisis of unprecedented scale. It has put huge pressure onto businesses and communities across Europe, as well as causing huge stress on public finances. But at the same time, it has confirmed that there are huge opportunities for Europe in building a resource-efficient society.
>
> (European Commission 2010, 2)

And the European Council insisted in 2011 that '[s]afe, secure, sustainable and affordable energy contributing to European competitiveness remains a priority for Europe' (European Council 2011, #2).

In fact, as the Commission pointed out in its *Roadmap for Moving to a Competitive Low Carbon Economy in 2050*, not pursuing the goal of reducing emissions now may turn out to be costly in the long run, as 'a less ambitious pathway could lock in carbon intensive investments, resulting in higher carbon prices later on and significantly higher overall costs over the entire period' (European Commission 2011, 5). On that account, climate action is not an emergency or even extraordinary measure anymore; it is part of the political calculus to optimise future performance and engage in economic steering. The ETS, for instance, in the eyes of the Commission, 'will be critical in driving a wide range of low carbon technologies into the market, so that the power sector itself can adapt its investment and operational strategies to changing energy prices and technology' (European Commission 2011, 6).

The lack of securitisations means that the articulations also make less justice claims. Instead, the negotiation of policies in the light of interests becomes core. At the same time, however, the desecuritisation of the EU discourse also leads to the shift of policies away from strict impartiality and towards a more flexible approach. In preparation of the COP in Cancun 2011, the European Council even made this explicit. After reiterating the need for 'a global and comprehensive legally binding framework', the heads of state and government now also announce that:

> In parallel with seeking an international agreement, the EU will also develop a more diversified approach to engaging with key partners in areas of mutual interest that help them reduce their emissions. In this context, the EU encourages regional initiatives to tackle climate change and promote green growth such as the recent Mediterranean Initiative on Climate Change.
>
> (European Council 2010b, #7)

Implicit in this statement is a concern for a plurality of actors within a procedural understanding of justice. Such a construction of justice, unlike a securitising move, makes no claim on behalf of a specific subject. It is a quiet invitation for dialogue, not a screaming accusation of injustice. It does not exclude that actor otherwise assume a position of superiority in their arguments, but in contrast to the impositions of presumed impartiality, such a more humble approach is no doubt more open to alternative conceptualisations of justice, which contributed, as we have seen in *Chapter 3*, to the forging of the PA.

4.5 The return of securitisation

The post-Copenhagen desecuritisation in EU articulations of climate change did not last long. While the Commission so far had stood out as a technocratic actor that largely steered clear of security talk, it became a strong securitising actor in the immediate run-up to Paris, perhaps to ensure the signing of an agreement. Later, in 2018/19, when it became clear that the Nationally Determined Contributions (NDCs) foreseen in the PA would not suffice to meet the two-degree, let alone the even more ambitious 1.5-degree target, and when the youth demonstrations spearheaded by FFF increasingly made the headlines, the European Council also returned to a securitising rhetoric – and even stronger than before.

However, at least as far as the Commission is concerned, securitisation now not only served the legitimisation of mitigation, but also of adaptation measures. In its 2013 EU Adaptation Strategy, thus, the Commission argued that:

> To avoid the most serious risks of climate change and in particular large-scale irreversible impacts, global warming must be limited to below 2°C above pre-industrial level. Climate change mitigation must therefore remain a priority for the global community. Whatever the warming scenarios, and however successful mitigation efforts prove to be, the impact of climate change will increase in the coming decades because of the delayed impacts of past and current greenhouse gas emissions. We therefore have no choice but to take adaptation measures to deal with the unavoidable climate impacts and their economic, environmental and social costs.
>
> (European Commission 2013b, 1)

This re-focussing on adaption also went in hand with an emphasis on Europe as a referent object:

> The temperature of the European land area over the last decade (2002–2011) has been on average 1.3°C above preindustrial level, meaning that the increase in Europe has been faster than the global average. Some extreme weather events have increased, with more frequent heat waves,

> forest fires and droughts in southern and central Europe. Heavier precipitation and flooding is projected in northern and north-eastern Europe, with an increased risk of coastal flooding and erosion. A rise in such events is likely to increase the magnitude of disasters, leading to significant economic losses, public health problems and deaths.
>
> (European Commission 2013b, 1)

Similar arguments are repeated in a 2018 Commission Communication, seeing climate change as 'magnifying the risks for instability in all forms' and noting that this 'would also have severe consequences on the productivity of Europe's economy, infrastructure, ability to produce food, public health, biodiversity and political stability' (European Commission 2018b, 1).

This is not to say that climate change was no longer seen as a global threat – it was, as a Commission Communication in preparation of the PA shows, which demanded urgent collective action 'with greater urgency and ambition' to avoid the worst consequences of a rapidly warming planet' (European Commission 2013a, 1). However, the prevailing securitisations, in combination with the stronger emphasis on economic interests in the Copenhagen-to-Paris phase, led to a focus on Europe as a referent object. Within Europe, justice was now sought for the particularly disadvantaged, who were especially vulnerable to the threat of climate change:

> Impacts vary across the EU depending on climate, geographic and socioeconomic conditions. All the countries in the EU are exposed to climate change ... However, some regions are more at risk than others. The Mediterranean basin, mountain areas, densely populated floodplains, coastal zones, outermost regions and the Arctic are particularly vulnerable. ... Climate change impacts are also expected to widen social differences across the EU. We need to give special attention to social groups and regions which are most exposed and already disadvantaged (e.g. through poor health, low income, inadequate housing, lack of mobility).
>
> (European Commission 2013b, 2–3)

This passage demonstrates very nicely the intertwinement of securitisation and justice claims. First, climate change is construed as an existential threat, then particular regions and groups as referent objects. Such speech acts establish specific subjectivities within the context of broader discourses (von Lucke 2020a, 15, 2020b). They are then taken to demand action, which is not explicitly linked to justice, but clearly implies that not taking action would be unjust towards those who the securitising move has established as the main referent objects.

After the immediate celebrations of the PA, as the limitations of the deal were becoming clearer, the Council returned to securitising moves in its conclusions. Interestingly, its documents now used a much more radical language than in the 1990s and early 2000s, performing textbook cases of such moves. In its preparations for the Katowice COP-24, for instance, the Council 'emphasise[d] the unprecedented urgency to step up global efforts to avoid dangerous effects of climate change' and was 'deeply concerned by the new evidence on the negative impacts of climate change', including the 'vulnerabilities, impacts and risks of further global warming to human societies and natural systems'. It was 'a matter of extreme urgency to strengthen the global response to the threat of climate change, sustainable development, and efforts to eradicate poverty' (Council of the European Union 2018, #1–#3).

Likewise, the Council Conclusions on Climate Diplomacy in February 2019 noted that:

> Climate change is a direct and existential threat, which will spare no country. The world is already witnessing multiple devastating impacts of climate change, yet action to stem it remains insufficient. ... In this context, it is a matter of extreme urgency to strengthen the global response to the threat of climate change in the context of sustainable development, and efforts to eradicate poverty. The Council is convinced that climate change acts as a threat multiplier and increasingly as a threat in its own right, with serious implications for peace and security across the globe. ... The Council is deeply concerned by the deterioration of global biodiversity, the world's water resources and ecosystems and the growing threats posed by land degradation, water scarcity, water-related risks and extreme water events.
>
> (Council of the European Union 2019, #1, #6, #7)

This language was repeated in the January 2020 Council Conclusions on Climate Diplomacy (Council of the European Union 2020, #1, #3). And even the European Council was now returning to securitising moves. In the conclusions of its October 2019 meeting, the European Council thus noted that the 'existential threat posed by climate change requires enhanced ambition and increased climate action by the EU and at global level' (European Council 2019a, #4). At COP-25 in Madrid in December 2019, the newly appointed European Council President Charles Michel talked of a 'climate emergency' and 'alarming trends' and used the rhetoric of being 'unconsciously "at war" with nature'. He asked his audience 'to dramatically change how we do things and revolutionise our approach' (European Council 2019b).

From a global justice perspective, there is an interesting difference between the securitising moves before Copenhagen and after Paris. Whereas in the early stages, the referent object had been mostly humanity and planet earth as such, post-Paris sees more differentiation. In particular, securitising moves focus on the most vulnerable populations either in poverty or particular geographic regions. The Council in its 2019 paper on climate diplomacy, for instance, emphasised the need to help 'the poorest and those in most fragile and vulnerable situations', which were 'the most exposed and the least able to respond or adapt to climate change' (Council of the European Union 2019, 3). At the same time, future generations figure less in EU documents, even though the public debate at the time was heavily influenced by Greta Thunberg and the FFF movement.

Thus, even the European Council referred to 'solidarity' and 'a social fair and just green transition' in its Conclusions (European Council 2019a, #4). The shift in the securitising moves had its clear corollary in a shifted focus in the call for action, which now was making justice claims not implicitly, but explicitly. This applied to global as well as EU-internal policy, which was supposed to 'ensure a cost-effective, just, as well as socially balanced and fair transition, taking into account different national circumstances in terms of starting points' (European Council 2019c, #4). This combined both substantive and procedural justice claims: to distributive justice in the request to ensure that the transition to a low-carbon economy is socially fair, and to non-domination with an opening towards mutual recognition in the demand to take into account different national circumstances.

4.5 Conclusion: different securitisations, different justices

In this chapter, we argued that the changes in the justice implications of EU global climate policy have been reflections of shifts in the securitisation of climate change within the EU discourse. We proposed three dimensions in which securitising moves could vary: in their referent object, their reference to immediate dangers or risk, and their time horizon. The analysis of European Council, Commission, Council, and Parliament documents showed that it is especially the referent object that mattered in EU climate policy. We thus demonstrated two moves.

First, after EU actors had heavily relied on securitising climate change, the documents we analysed showed a process of desecuritisation especially in the immediate aftermath of the Copenhagen COP-15 in 2009. EU actors moved from explicitly invoking climate change as a threat to leaving that threat implicit and focussing on the development of an appropriate policy to reduce GHG emissions. This went along with an increasing linkage of climate change with energy security and economic stability, in which interest, not justice was central. Council conclusions and other documents became more detailed and managerial, and even the recitals at the beginning of official documents steered clear of climate security claims.

Second, from 2013 onwards and especially after the implementation of the PA had started, we witnessed a return to open securitisation moves, which now, however, differentiated the impact on different referent objects. Whereas in the late 1990s and early 2000s, articulations of climate change referred to its global impact on humanity and planet earth, and often included an explicit future-oriented dimension, they now particularly focussed on the effects on specific vulnerable groups in the present or near future. In this context, justice claims that arose from these securitising moves were made more explicit and shifted from a focus on impartiality and responsibilities of mitigation to a focus on non-domination and mutual recognition as well as an increasing emphasis on adaptation measures and distributive justice.

These moves broadly tie into our analysis of EU climate policy and its justice implications, which we presented in the preceding chapter. Our argument is not that there is a sort of explanatory logic to this linkage. Instead, we wanted to demonstrate how justice claims, in their reasoning, go along with securitising moves that construct the referent objects at their center. These constructions of climate change as a threat and its referent objects are politically contested as they lay the foundations of rights and duties, of policy proposals and funding streams. As such, it matters a great deal whether climate change is seen to primarily threaten Europeans or vulnerable communities in the Global South, the economy or future generations, a presumed universal community modelled on a Western standard or a diversity of communities in different locales. These controversies are reflected in the engagement between the EU and developing countries, to which *Chapter 5* now turns.

5 The EU and global climate justice seen from the outside

The existing literature on the European Union (EU)'s role in international climate governance rarely discusses other countries' perceptions of the EU or draws on insights from the growing literature on developing countries in the climate negotiations (e.g. Najam 2005; Qi 2011; Hochstetler and Milkoreit 2014). This chapter focusses on how outsiders perceive and interpret the EU in the climate regime. This change in perspective helps us to critically evaluate the EU's climate strategy and its focus on justice as impartiality, and discuss in more detail the value of non-domination and mutual recognition when it comes to recognising and empowering actors from the Global South. These discussions alert us to the structural injustices of the historically grown international order and demonstrate that the EU's claims to universality are often founded on self-centred perceptions of what is right and what is wrong. Thus, the EU should take mutual recognition more seriously in its engagement in the global climate regime, although, in a state-based order, elements of mutual recognition will often be mixed with elements of non-domination.

As a group of large developing countries, Brazil, South Africa, India, and China (the BASIC countries) have played a central role in international climate negotiations since the establishment of the United Nations Framework Convention on Climate Change (UNFCCC) in 1992. Although culturally, economically, and politically different, these four countries operated together within the Group of 77 (G77) from the start of the climate negotiations, and have since 2009 maintained a formal cooperative alliance, BASIC, in the negotiations (Kasa, Gullberg, and Heggelund 2008; Qi 2011). As prominent representatives of developing countries and responsible for an increasing share of global Greenhouse Gas (GHG) emissions, the BASIC countries have received increasing attention from both scholars and climate negotiators over the past 20 years (e.g. Hurrell and Sengupta 2012a; Papa and Gleason 2012). Because the BASIC countries are among the world's largest economies, there has been some debate as to whether they should still be classified as 'developing countries' or rather constitute 'emerging economies or powers' (Qi 2011; Hurrell and Sengupta 2012a; Hochstetler and Milkoreit 2014; Michaelowa and Michaelowa 2015). Apart from academic disputes, the classification as developed, developing or emerging country is also a highly relevant question when it comes to the climate negotiations because of the differences in emission reduction duties associated with these categories. The BASIC countries refer to themselves as 'developing' and are somewhat uncomfortable with the 'emerging' label. For the sake of simplicity, we will use the term 'BASIC countries' or 'large developing countries' in the following.

This chapter pays particular attention to the relationship between the EU and the BASIC countries in the climate negotiations, focussing on how these countries understand the negotiations and the EU's role in them. In analysing the EU from 'the outside', this chapter does not primarily aim to make normative evaluations of the actors' justice perspectives or make arguments of what is more just or fair, although we will return to such an evaluation in the Conclusion. Rather, we here follow David E. Storey's suggestion that

> [...]given the failure of ethics to penetrate climate policy, and of climate policy to gain political traction, perhaps we ought to stop asking whether our accounts of climate justice are 'right or wrong' and start asking why they have not worked so far.
>
> (2019, 38)

The chapter is based on qualitative, empirical process tracing with a triangulation of data sources (George and Bennett 2005). First, we have collected written primary and secondary sources such as official documents, statements, press releases, letters, speeches, newspaper articles, reports, and published research. Second, we have collected data through fieldwork in Brazil in 2013 and 2014, and in India in 2014, and participation and observation at the Rio Clima conference in 2013, the Delhi Sustainable Development Summit in 2015, and the Conference of the Parties to the United Nations Framework Convention on Climate Change 23 (COP-23) in 2017. Third, this chapter builds on 22 semi-structured interviews with international experts, negotiators, Non-Governmental Organisation (NGO) representatives, politicians, bureaucrats, businesspersons, consultants, and scientists from the EU and the BASIC countries between 2013 and 2019.

5.1 Historical perceptions of climate change and their justice implications

From the start of the climate regime, the EU and many developing countries had very different perceptions of the climate problem and its inherent questions of global justice (Miller 2004). To developed, and especially European countries, climate change was a global *scientific* challenge and global commons problem, and what was needed to solve it was more accurate scientific information, through unifying and improving global knowledge about climate change and how it could be addressed (Miller 2004; Lahn and

Sundqvist 2017). This scientific understanding lay behind the establishment of the Intergovernmental Panel on Climate Change (IPCC) and played a key role in the establishment of the Kyoto Protocol and the development of EU climate policies. To developing countries, represented by the G77, climate change was primarily a *political* challenge related to historically unequal dependency structures in the international order, developed countries' overuse of resources for decades, and challenges of 'poverty, development, equity, and access to technological and financial resources' (Miller 2004, 59; see also Najam, Huq, and Sokona 2003; Aamodt 2018a). Developed and developing countries agreed that mitigating climate change was the main *solution* but disagreed on defining the *problem* to be solved in the climate negotiations, and thus on who should bear the cost of mitigation. Arguing that the IPCC was too dominated by developed countries, developing countries opposed the initial idea of giving the IPCC a major role in climate negotiations (Miller 2004). They insisted on a new multilateral body for climate negotiations based on the United Nations (UN) principle of 'one country, one vote', and thus the UNFCCC was established (Miller 2004).

Although issues of sustainable development and equitable burden-sharing were included in the UNFCCC from the start, the G77 argued that those issues were neglected in the subsequent negotiations and in the Kyoto Protocol (Najam, Huq, and Sokona 2003). As *Chapter 3* exemplifies, the Kyoto Protocol with its internationally agreed mitigation targets, timetables, and sanctions was very much in line with the EU's scientific understanding of the climate problem and its preference for legally binding, supranationally enforced emission reduction targets. In this context, the EU has focused on countries' current GHG emissions and justice for future generations, arguing that science tells all countries to act now to avoid permanent damage to the planet (Lahn and Sundqvist 2017). According to this scientific understanding, the fast-growing GHG emissions in large developing countries are just as important to mitigate as the stabilised, although high, GHG emissions in developed countries.

In contrast, in the BASIC countries' *political* understanding of the climate problem, the historical responsibility for accumulation of GHGs in the atmosphere needs to be the starting point for discussing substantial climate justice (La Rovere, Macedo, and Baumert 2002; Torney 2015a; Aamodt 2018a). Instead of focussing mainly on who should mitigate and how much, this political understanding focusses on who should *pay* for mitigation (Warlenius 2018) (see also the discussion in *Chapter 2* on assigning responsibility and allocating costs).

To discuss how these two different understandings of the climate problem relate to the perceptions of justice as impartiality, mutual recognition, or non-domination, it is necessary to pay some attention to how the BASIC countries and other developing countries perceive the EU as an actor in the climate regime. These countries' understanding of the international order and Europe's role in this order is decisive for their understanding of procedural climate justice. Thus, while the EU sees itself as a leader in the climate negotiations by virtue of its supposedly ambitious climate policies and high engagement in the negotiations, this self-understanding is not always matched in other actors' perceptions of the EU. The normative power Europe literature argues that the sui generis of the EU makes the union able to be a different kind of international power (Manners 2002). Outside Europe, however, the EU's uniqueness is not necessarily seen as increasing its power but often rather complicates the recognition of the EU as an independent actor (Laatikainen and Smith 2006). This is reinforced by the fact that although the EU negotiates as a block in the climate regime, this is not the case in many other international regimes. Even in the UNFCCC, EU member states send additional delegations to the negotiations. Thus, the EU is mainly understood as an institutionalisation of cooperation between European states, and rarely as an actor that has power or autonomy in addition to the power and autonomy of the member states (Interviews 2014a, 2018a, 2019a).

Beyond this different perspective on EU leadership, to many developing countries Europe primarily represents the former colonial powers (Roberts and Parks 2007; Sierp 2020). The colonial legacy is still strong in developing countries and developing countries' societal structures and political systems are often hybrids between national and European customs and rules. Whereas the European integration project has been rather inward-looking, without much need for an understanding of the world outside Europe, developing countries' understanding of their identity is always somewhat related to what they are not: European. As López-Alves (2012) argues regarding the domestic–international interlink in Latin American countries, the international always had a strong influence on modern nation-building in developing countries. This dependency is part of national self-understanding, and the modern nations were built in contrast to the imposed European rule (López-Alves 2012). As discussed in the *introduction* and in *Chapter 3*, the literature on the EU as a normative power pays little attention to the EU's colonial legacy and assumes that the EU aims to promote universal norms of justice and human rights both inside and outside Europe, and that its ability to enforce these values and norms in Europe gives it credibility and legitimacy internationally (Manners 2006).

However, whereas the EU member states have been rather successful in drawing a line under former disputes in their integration project, such historical closure has not happened in most of the former colonies. Instead, developing countries' collective memory of relations with European countries and struggles for independence starts centuries ago. The current post-colonial national identities are constructed on the memory of past oppression. Military, economic, or normative power demonstrations from Europe are thus often understood in the discourse of European imperialism.

Within the EU there is surprisingly little self-reflection and recognition of the importance of Europe's colonial past in current international relations (Sierp 2020). As Sierp argues

> [t]he EU remains curiously quiet about the memories of imperialism and colonialism. Humanitarian catastrophes, civil wars and border conflicts, state collapses, terror attacks and environmental and climate catastrophes appear to the European public to have little to do with this history.
>
> (2020, 688)

In relation to the Global South, the EU is therefore in a particularly unfortunate position for having normative power: trying to 'shape conceptions of "normal"' (Manners 2002, 239) will very easily be regarded as a European imperialist endeavour.

The EU's preference for binding emissions reduction targets, supranational control, and efforts by all countries is directly linked to the scientific understanding of the climate problem and the union's inability or unwillingness to adequately deal with the colonial legacies of many of its member states (Aamodt and Boasson 2020). Climate science clearly shows that it does not matter where GHGs are emitted or mitigated, the global effect will be the same (IPCC 2014). In light of this knowledge it is logical to prefer a climate deal that includes all emissions, prevents free-riding, and has a good system for monitoring and verifying the emissions and mitigation efforts of all the parties. A regime where all are included and controlled by international bindingness is the closest one can come to effective international governance of a global commons problem like climate change (e.g. Hovi, Skodvin, and Aakre 2013). As outlined in *Chapter 3* such a solution is perceived as the most fair and effective by the EU, because the international control mechanisms can protect the individuals most vulnerable to effects of climate change by providing the most efficient, and therefore understood as most ambitious, regime. In a functioning impartial system, no one is left (or allowed to lag) behind.

Nevertheless, as we have already discussed in *Chapter 2*, an impartial climate regime requires finding 'some sort of neutral, unbiased solution [...] justifiable from the perspective of all', and this solution needs to be enforced by 'legal instruments beyond the state' (Sjursen 2017, 8). Thus, before even getting to the nitty-gritty details of deciding targets and timetables, an impartial agreement requires that the parties believe in the possibility of neutral supranational authority. To be legitimate, such an authority must enforce rules and regulations that are perceived as fair and equal. The colonial period and the subsequent telling hierarchy of the First, Second, and Third World division, have created a common experience among developing countries of being treated like 'the others' in international politics (Tickner 2003; Roberts and Parks 2007; Acharya and Buzan 2010). These countries simply do not think the current international order is constructed to treat parties equally, or able to do so in practice (Roberts and Parks 2007). Thus, they seriously question the possibility of neutral (impartial) solutions within this order.

This sentiment is not limited to the climate regime. Especially large developing countries like China, Brazil, or India indeed want a stable international order, but their perception of stability is different from what most Western International Relations (IR) scholars assume (Stuenkel 2016). According to Stuenkel, large developing countries '[...] regard multipolarization and the contestation of the US hegemony as a necessary change to assure that rules and norms will be respected' in international relations (2016, 359). With regards to climate change, the BASIC countries do not believe they will be treated fairly by an internationally binding climate regime and instead prefer a regime without supranational authority (Roberts and Parks 2007; Interviews 2014b, 2014d, 2017g; Zhang 2016). Among the justice conceptions discussed in this book, the BASIC countries often argue in line with a state-centric understanding of justice, thus mainly defending non-domination. However, they also emphasise structural inequalities that exceed the territorial state order defended by non-domination. Although non-domination recognises power inequalities between states and argues that the international order needs to have measures to prevent misuse of power and protect less powerful states, mutual recognition also includes recognition of unintended and often intangible power abuse by the current order's privileged actors, including non-state actors such as Transnational Companies (TNCs).

In such a perspective, recognition goes beyond state sovereignty towards a focus on autonomy. As Tickner argues, '[r]ather than being rooted in juridical notions of sovereignty, [autonomy] is a markedly political concept, and is viewed as an instrumental tool for safeguarding against the most noxious effects of the international system' (2003, 319). Autonomy requires active recognition between parties, meaning that those historically privileged by the international order (the EU in our case) recognises the subjectivity of those historically objectified (the Global South in our case). According to Suiseeya, such 'recognitional justice [...] requires that *differences* in cultures, lifeways, and ways of knowing are recognized, respected, and appropriately incorporated in environmental policy processes from conception to design, implementation, monitoring, and evaluation' (2016, 3, our emphasis). Justice as mutual recognition thus 'requires scholars and policy makers to identify and confront the ways in and pathways through which cultural norms infuse and shape environmental institutions to perpetuate the dominance of these norms' (Suiseeya 2016, 3). The European climate strategy, providing the building blocks of EU 'leadership by example', may not be as universal as it appears to European practitioners and experts. As Stuenkel notes in his discussion of 'soft power' as an analytical concept, 'Western analysts often confidently assume that China will never be as attractive as democratic regimes, yet populations in Africa or other parts of the world may not necessarily agree' (2016, 356). As an extension of this, Zhang (2016) argues that from a Chinese perspective the 'liberal order' is an extension of the western colonial dominance. And further, that the liberal order's

> division of the world into a liberal zone of law, order and justice and a non-liberal zone of violence, injustice and chaos is seen not just as an emerging reality, but a morally defensible proposition that justifies and legitimates an extraordinary range of interventionist and intrusive or otherwise coercive activities in the non-liberal zone.
>
> (Zhang 2016, 804)

The EU's framing of its preference for a supranational and binding regime as universal and in the interest of vulnerable individuals, may thus be perceived as an imposition, an attempt to dictate global governance rules while hiding behind the vulnerable.

The EU's well-intentioned 'leadership by example' thus does not go well with its reluctance to accept the importance of the colonial legacy in international relations. The EU's defence of the most vulnerable, especially the Least Developed Countries' (LDCs) right to growth and development and the support of Common But Differentiated Responsibilities (CBDR), was welcomed and appreciated by developing countries from the start of the climate negotiations. But the weak involvement of these countries in the Kyoto negotiations gave them the impression of 'being counted, but not count'[1] when core issues of the climate regime were decided (Roberts and Parks 2007). The duality in the EU-developing countries relations was also demonstrated at COP-8 in New Delhi in 2002, when a massive alliance of Global South NGOs and indigenous peoples, supported by G77, united under the slogan 'climate change is a human rights issue' (Roberts and Parks 2007). This created a reaction among European NGOs and EU negotiators that worried 'their hard-won treaty to address and reverse climate change [would be] transformed into one on Third World development' (Roberts and Parks 2007, 134).

The BASIC countries and the G77 have since the beginning of the UNFCCC insisted on the developed countries' historical responsibility for climate change, not only as a basis for calculating mitigation targets, but also as a plea for the developed countries' self-recognition of their privileged position as the historical beneficiaries of the international order (BASIC 2013; Gupta and Chaudhary 2014; Saran 2015). As outlined thoroughly in *Chapters 2* and *3*, the CBDR principle mostly strengthens justice as mutual recognition. In the UNFCCC CBDRs is followed by 'and Respective Capabilities' (RCs). CBDR-RC forms the cornerstone in the BASIC countries' arguments against a centralised, top-down, binding agreement. To the BASIC countries CBDR recognises different parties' different starting points and historical responsibilities. In an autonomy context, the RC is important as it points to each party's ability to self-determine its own mitigation and adaptation capacity, and the other parties' trust and recognition of that ability. In recognising autonomous decisions, one shows and generates trust, while the EU's preferred approach, in its zeal to prevent free riding, has embedded mistrust.

However, the BASIC countries are also often seen to be inconsistent because they want binding commitments for developed countries, but not for developing countries like themselves that also have high emissions (Hurrell and Sengupta 2012a; Hochstetler and Milkoreit 2015; Rajamani 2017). The BASIC countries' main argument in this regard is that the developed countries have not lived up to the obligations they already have under the UNFCCC and Kyoto Protocol, while the developing countries are already mitigating more than they are committed to (Zuma 2009; BASIC 2013; Ananthakrishnan 2015). They emphasise *differentiation* in the climate regime, but usually only differentiate *between* the groups of developed and developing countries, and not *within* the groups, and especially not within countries. As argued by India's former special envoy on climate change,

> [t]here is a difference between the emissions of developing countries which are "survival" emissions and those of developed countries which are in the nature of "lifestyle" emissions. They do not belong to the same category and cannot be treated on a par.
>
> (Saran 2015)

This perspective comes very close to Shue's differentiation between 'subsistence' and 'luxury' emissions in his discussion of climate justice approaches, see *Chapter 2* (Shue 1993). The emphasis on differentiation is thus placed somewhere between mutual recognition and non-domination. The logic is that the developed countries had the chance to make responsible, sovereign mitigation decisions, but did not use that chance, and therefore need to be treated differently within the regime to ensure the safety of vulnerable countries and people. The EU's performance in relation to other developed countries is seldom recognised, and if so, still regarded insufficient (Interview 2013a; Hochstetler and Milkoreit 2015).

5.2 The global south and the EU's role in climate negotiations: lost in translation?

As already touched upon in *Chapter 3*, a key explanation for the EU's limited influence during the 2009 climate summit in Copenhagen was its unwillingness and inability to address recognitional justice in the climate regime. Already at the COP in Bali in 2007 it had become clear that the BASIC countries disagreed with the EU's aim of a new agreement that was binding for all countries (Groen 2016). Instead, in line with their aim of autonomy, the BASIC countries argued in favour of Nationally Appropriate Mitigation

Actions (NAMAs) in Bali (Qi 2011) and were for the first time willing to discuss voluntary commitments for developing countries (Li 2017). However, both the BASIC countries and the G77 underlined the idea that the distinction between developed and developing countries should remain, and insisted on the Annex-1 countries fulfilling their binding Kyoto commitments (Earth Negotiations Bulletin 2007). Supported by the G77, the BASIC countries argued that the Bali text should continue to differentiate between developed and developing countries, making the latter's commitments dependent on sustainable development and RCs (Earth Negotiations Bulletin 2007; Qi 2011). Although this suggestion was not in line with the EU's preference for a binding agreement and heightened responsibilities for large developing countries, the EU decided to support the BASIC countries' suggestion in order to reach agreement on a text in Bali (Qi 2011).

This experience in Bali should have made it clear to the EU that the large developing countries were increasing both their own willingness to contribute to global climate governance and their efforts to craft a new climate agreement, in accordance with their preference for autonomy and differentiation. However, in the preparation for Copenhagen, the EU engaged remarkably little in climate dialogues with developing countries, especially when it came to trying to understand other parties' positions (Aamodt and Boasson 2020). Any views that questioned the EU's preferred type of agreement were dismissed as less ambitious, and the EU's climate diplomacy's main aim was to communicate the EU's solution (Aamodt and Boasson 2020). This lack of mutual recognition is not new, and the EU has often been criticised for 'talking at' instead of 'talking with' in diplomatic settings (Chaban and Knodt 2015). It also confirms the developing countries' understanding of the world as structurally unequal, where the privileged are seemingly unaware of (and often deny) their privilege and insist on all parties being treated equally.

An example in this regard is the EU's attempt to ally with the LDCs to split the G77 and put pressure on the large developing countries before Copenhagen. While being praised by some EU negotiators as an effective negotiation strategy (Interviews 2018c, 2018d) the BASIC countries saw this strategy as a confirmation of the Annex-1 countries' hidden agenda of making the developing countries responsible for mitigation in order to pay less themselves (Abranches 2010). They thus perceived it as deepening mistrust between developed and developing countries (Dasgupta 2009; Abranches 2010). They were not convinced by the EU's claim of wanting to take care of the interests of the most vulnerable, the LDCs, and future generations.

Moreover, even the countries and people that the EU claimed to champion – the poorest and most vulnerable – had a differentiated perspective when it came to distributing the burdens for mitigation. On the one hand, the negotiation alliances of Small Island Development States (SIDS) and LDCs demanded high mitigation efforts also from the BASIC countries (Pasternack 2009; Hallding et al. 2011). They also disagreed with the BASIC countries on binding commitments. They perceived climate change as an immediate danger – and thus similarly to the EU engaged in clear securitisation moves (see also *Chapter 4*) (Oels and von Lucke 2015, 44) – and wanted all countries to be legally obliged to act (Pasternack 2009). In addition, groups within the BASIC countries pushed for more climate action. For instance, Indian NGOs openly criticised that their government's emphasis on equity in the international climate negotiations was not matched with a focus on equity in domestic politics (Lele 2012), and Brazilian NGOs and researchers have allied with the British Embassy to suggest strong mitigation measures for Brazil (Biderman 2009; Aamodt 2018b). On the other hand, these vulnerable actors agree with the BASIC countries regarding the EU's and other rich countries' historical responsibility for mitigation. As argued by an Indian civil society activist when asked about the EU's role in Copenhagen: 'We used to have a lot of respect for the EU, they provided leadership although they were not taking their fair share of the burden. 20-20-20 is nothing' (Interview 2017h). As a consequence, the EU's strategy of splitting the G77 by seeking an alliance with the LDCs failed as a strategy for pushing the BASIC countries to accept a binding agreement (Abranches 2010).

Another event demonstrating the diverging perceptions of the EU and the BASIC countries in Copenhagen was the debate about a leaked negotiation document – the so-called 'Danish text'. The negotiations text was written by Denmark in cooperation with some other (mainly developed) countries and leaked to the newspaper *The Guardian* before it had been presented to all UNFCCC parties. The Danish text was dominated by a scientific understanding of the climate problem, and developing countries saw its suggestions as favourable to the developed countries, fuelling their suspicion that Europe and the United States (US) were brokering a deal outside the multilateral negotiations (Vidal 2009; Abranches 2010). The Danish text did not deviate much from the positions already communicated by the COP presidency, and the Executive Secretary of the UNFCCC, Yvo de Boer, said the text was rather balanced, but also the main cause of the Copenhagen failure (Vidal 2010). The BASIC countries reacted strongly and negatively to the text for two reasons. First, because the text was written by some selected parties and not as a multilateral process. Such a procedure is not very unusual in the negotiations, but with the already-existing tensions between developed and developing countries, several developing countries protested strongly when the text was leaked before it was formally presented (Vidal 2009). Second, the text suggested differentiation within the group of non-Annex 1 parties. This suggestion was seen as an attempt by the developed countries to flee from their mitigation responsibilities and pass them on to the large developing countries (Abranches 2010). The text thus echoed the EU's argumentation in favour of a breach with the Annex 1/non-Annex 1 divide, arguing that the 'emerging economies' should have more demanding mitigation commitments (closer to the Annex 1 countries' commitments under Kyoto) than the LDCs should have (Council of the EU

2009). In terms of both numbers and fairness, this argument should not be that controversial. Looking at the distribution of states' historical, current, or per capita emissions, it is clear that the LDCs are the ones least responsible for climate change, and the ones least able to mitigate. However, the argument was politically and diplomatically controversial. From the BASIC countries' perspective, it implied a renegotiation of terms already agreed under the UNFCCC; the dichotomous division between developed and developing countries (Qi 2011). The EU's suggested targets for different parties in Copenhagen were based on an IPCC report and argued to be neutral (Lahn and Sundqvist 2017), but they nevertheless sent two unintended signals to the BASIC countries.

The first signal was that the EU was not aware of or did not recognise the recent changes in domestic climate policies in all the BASIC countries. China had launched its National Climate Change Programme in 2007, India had adopted its economy wide National Action Plan on Climate Change in 2008, Brazil had adopted a climate law in 2009, and South Africa was well in the process of adopting national climate policies (Aamodt and Stensdal 2017; Tyler 2010). Had the EU been less narrowly focussed on scientifically derived targets and timetables, and more conscious of the requirements of reciprocal diplomatic dialogue, the BASIC countries' trust in the EU's stated intentions in Copenhagen would have been higher (Aamodt and Boasson 2020).

The second signal sent by the EU's suggested targets was that the Union did not trust other countries' ability and willingness to make voluntary commitments in line with their own capabilities. Interviewees highlight that the EU's climate diplomacy has, especially before Copenhagen, tended to be a one-way communication where the EU suggests solutions or policies to be adopted in developing countries, following the tradition of talking at and not with (Chaban and Knodt 2015; Interviews 2013c, 2014a, 2019b). Where the BASIC countries have asked for climate *cooperation*, the EU has tended towards *aid*. The EU's aid strategy worked well in smaller and less-developed countries, and also in specific bilateral projects in the BASIC countries. However, being perceived to have an attitude of being the one that can teach others, but having nothing to learn oneself, works poorly in diplomatic settings.

Keeping the divide between developed and developing countries in a new climate deal was one of the main reasons for formalising the BASIC cooperation before Copenhagen (Michaelowa and Michaelowa 2015). At the end of November 2009 all the BASIC countries met in Beijing and communicated unity on several aspects regarding the upcoming negotiations in Copenhagen (Dasgupta 2009). The invitation of the Chair of G77 to this first BASIC meeting shows the BASIC countries' strong mutual interest in avoiding a divide within the G77 during COP-15. In Beijing, both the BASIC countries and the G77 stated that they would not let the developed countries force their conditions on developing countries, and that they would contribute to reaching consensus in Copenhagen (Dasgupta 2009).

In the *scientific* understanding of climate change, emission levels are the main indicator for the annex 1/non-annex 1 division. Because global emissions in 2009 were divided quite differently between countries compared to the situation in 1992, the EU, supported by the US, argued it was time to reconsider the dichotomy (Wu 2012; Bäckstrand and Elgström 2013). China was a particularly important example; China's emissions had almost tripled since 1992 (Wu 2012). By contrast, in the *political* understanding of climate change the definition of a developing country is historical, and rising emissions do not change a country's history. Although China became the world's largest emitter in 2007, China was ranked as number 92 on the UN's list of Human Development Indicators, with medium human development in 2009 (UNDP 2009). Chinese negotiators argued that continued economic development was pivotal to the Chinese population, and that China's emissions were 'subsistence' emissions in comparison to the excess 'luxury' emissions in developed countries (Kopra 2012). To the Chinese, the developing country status was an issue of sovereignty. The State Council's statement that China is a developing country was not for other countries to question (Interview 2017g). India also reacted strongly to the suggestion to renegotiate the Annexes, arguing that the focus on emerging economies was not relevant for the climate negotiations:

> This comes from Europe, I mean the developing countries, we did not invent the emerging economies label. We were artificially labelled as emerging economies, we were pushed by the Western countries, pushed into BASIC. Maybe the label has a meaning when it comes to trade and economy, but there is no justification in the climate context.
>
> (Interview 2014a)

A veteran in Indian climate policymaking referred to the fable about the sun and the wind when explaining why the EU kept failing to lead in the negotiations (Interview 2014d). In the fable the sun and the wind compete to get a man to take his coat off. Although the wind uses a lot of strength, the man just starts to hold even harder onto his coat. The sun, however, is able to make the man take off his coat with little effort. In climate dialogues with developing countries before and during the Copenhagen COP, the EU's attempts to push for ambition were perceived as equivalent to the wind's. As outlined in *Chapter 3*, instead of just presenting own targets and ambitions in Copenhagen, the EU portrayed its own targets as the most ambitious, and partly conditioned them on other countries' willingness to make binding commitments. This strategy was needed for internal-EU political reasons, but it came out as arrogant to the BASIC countries that had presented quite ambitious voluntary targets in Copenhagen. In the end, as expressed by another Indian interviewee: 'Copenhagen was a complete breach of trust' (Interview 2017f). The BASIC countries perceived that the EU did not trust their honest intention to contribute to global climate change mitigation through voluntary national commitments, and the EU's honest intention to shield the LDCs from climate demands, was perceived as a strategy to split developing countries and reduce own commitments.

5.3 The global south and the road to Paris: leaving impartiality

As discussed in *Chapter 3*, the perceived failure in Copenhagen in combination with the establishment of the European External Action Service (EEAS) in 2010, led to changes in the EU's climate diplomacy (Torney and Cross 2018; Aamodt and Boasson 2020). Yet, when asked about the EU's climate diplomacy between the COPs, BASIC country interviewees do not mention the EU as a key diplomatic actor (Interviews 2017f, 2019a, 2019d). Most of the EU's climate diplomacy work both before and after Copenhagen was done by member state embassies (Torney and Cross 2018). Nonetheless, the increased role of the Green Diplomacy Network (GDN), and the enhanced focus on diplomacy and the need to *understand* other parties, not just inform them, was noticed by BASIC countries (Aamodt and Boasson 2020).

The perceived failure in Copenhagen also paved the way for new actors among the developing countries to take a more prominent role in the negotiations. Negotiators from both developed and developing countries were impressed by the diplomatic strength of the Mexican hosts of COP-16 in Cancun in 2010 (Qi 2011; Walker 2018). The Cancun COP showed how dialogue with an emphasis on process, thus a strengthening of mutual recognition and non-domination, was perceived as more just by most parties than the EU's emphasis on targets and 'one right solution' had been in 2009 (Walker 2018).

After the launch of the BASIC alliance in Copenhagen, the four countries took rather different roles in the subsequent negotiations. China and India were the ones most clearly offended by the claim that the BASIC countries were to blame for the failure in Copenhagen (Torney 2015a), and in India it caused the Parliament to strongly criticise the environmental minister for being too Western-friendly (Dubash 2012). Brazil has a more pragmatic diplomatic tradition and kept working on its negotiation proposals in confidence that deforestation rates were declining, and Brazil was reducing emissions more rapidly than most other countries, including the EU (Hochstetler and Viola 2012; Howard 2014). After Cancun, South Africa started preparing for COP-17 in Durban in 2011, and therefore held a low profile in the most heated debates. At COP-17, Brazil and South Africa stayed out of India's hard-line arguments with the EU on the legally bindingness of a new climate agreement (Rajamani 2017).

Of the BASIC countries, India is the least developed, and around 300 million Indians, equivalent to two thirds of the EU population, lack access to modern energy services (Dubash et al. 2018). India currently has low per capita emissions (about two tons per year as compared to seven tons in the EU), but also an acute need for development in terms of infrastructure, food, and energy security. India's government estimates that half of the India of 2030 is yet to be built (Government of India 2015). Although the other BASIC countries also strongly disagree with the argument of binding emissions for emerging economies, the EU's justification of this argument is most misplaced in the Indian context. In the Annex 1 countries, most GHG emission sources are known and emission trajectories can be fairly well modelled by making changes in known parameters. An important reason behind the BASIC countries' rejection of a binding agreement is that they know much less about their future emissions. Being embedded in a cultural and diplomatic tradition where it is important not to lose face, and where giving one's word is regarded to be binding, it is particularly difficult for India and China to make binding emissions commitments. These countries are still in the process of building systems that would give them a solid and reliable overview of the emissions they already have, and compared to the EU's situation, it is much riskier for them to launch their ambition tied to a fixed number. They do not know for sure that the numbers included in the equation in the first place are correct.

Comparing ambitions is also misleading in this context because the announcement of targets has very different political traditions in different countries. In the EU, targets are aspirational, they are set after negotiations between political actors – though partly based on scientific projections and influenced by other actors such as industry lobbyist or civil society activists – where some always want a higher target (we should do more) and some want a lower target (is it expensive to deviate from status quo) (Boasson and Wettestad 2016). In the BASIC countries, targets are usually set as the minimum, being the level one is quite certain to reach with currently adopted and implemented policies (Aamodt and Stensdal 2017). There are domestic political debates on climate measures and policies to be adopted and implemented, but discussions of economy-wide targets are usually not part of these debates (Aamodt and Stensdal 2017). Brazil's target from 2009, reducing emissions between 36.1 and 38.9 per cent below business-as-usual trajectories by 2020, is an example of a target carefully calculated based on the government's expected emissions trajectory with already adopted policies: 'nada novo!' ('nothing new!'), as a climate advisor in the Brazilian senate expressed it (Interview 2013b). Similarly, South Africa's Intended Nationally Determined Contributions (INDCs) presented in 2015 'builds on a strong basis of existing national policies and actions' (Zuma 2015). The EU's targets are of course not random, but they are set in a more political way than the technocratic targets of the BASIC countries, and the insistence on them as being ambitious (the *most* ambitious) is perceived as arrogant from the outside. As argued by a Chinese interviewee: 'It is difficult to compare. All use different parameters and measures, and methodologies. We need to respect each country's decision' (Interview 2017g). The change in the negotiations from globally agreed, binding targets to INDCs after Copenhagen also enabled the EU to see the negotiations as a part of a larger policy developing process (with domestic policymaking centre stage), and not a competition about numbers.

The increased engagement of Central and Eastern European countries in EU climate policymaking increased the Union's climate policy challenges not only internally, but also internationally, at least in regard of being perceived as a unitary actor and leader. According to several interviewees in large developing countries, the internal opposition from the Eastern European countries and the fossil fuel-based industries to strong climate measures signals that the EU is nothing out of the ordinary regarding climate action (Interviews 2014a, 2017h). Most other countries also have regions and cities that are national climate leaders, while other regions and cities are laggards. The EU's power in relation to the member states is unclear to the BASIC country governments (Interviews 2017f, 2018a, 2018m), and several mention that it is only the North-Western member states that have climate ambitions, while the Southern and Eastern member states are burdensome to the Brussels-based negotiators (Interviews 2014a, 2017h, 2019a, 2019c). As one Brazilian climate policy expert puts it 'Italy is not Sweden. The whole world knows this' (Interview 2019a). Although this impression weakens the EU's image of being a climate leader, it can also be positive for the EU's ability to engage in recognitional climate dialogue with developing countries.

Compared to the EU, individual EU member states have larger abilities to engage in projects of technological cooperation, research collaboration, and development that creates a reciprocal environment of learning between partners (Interviews 2014c, 2019d). BASIC country interviewees mention several of these projects as important for dialogue and trust building. Some examples are the United Kingdom's (UK) large amount of climate related collaboration projects in Brazil, cooperation between Denmark and China on wind power, and the International Solar Alliance launched by India and France (Interviews 2017f, 2017g, 2019a). The diplomatic efforts of Poland and France as COP hosts (in 2013 and 2015, respectively), including high-level visits to India, was positive for India's perception of the UNFCCC process and the upcoming Paris negotiations (Interview 2017f). Thus, in creating trust between negotiation parties, the EU's plurality is often a more important asset than its unity. France's ability to distance itself from the EU before and during the Paris negotiations was a factor that contributed to the perceived success in Paris (Walker 2018; Raman n.d.). In bilateral cooperation, the EU member states are also perceived to have more flexibility and can contribute positively as a partner in concrete projects (Interviews 2017f, 2017g, 2019a).

As *Chapter 3* has pointed out, the Paris Agreement (PA) is very different from the kind of deal the EU tried to get in Copenhagen. Besides the changed strategy of the EU, certain red lines drawn by the US and China played an important role for the development of the novel regime architecture in Paris (Interviews 2019a, 2019b, 2019d). Their joint announcement of climate targets in 2014 demonstrated their willingness to contribute to a new climate deal in Paris, but also drew the limits of that deal quite clearly. The EU had already adapted their negotiation strategy and was willing to accept a higher degree of voluntariness (non-domination) and difference (mutual recognition), without giving up entirely on bindingness and top-down elements. Thus, during the Paris negotiations, the EU at first argued that although the NDC targets and measures were decided domestically, the attainment of the targets should be legally binding at the international level (Rajamani 2016). This was supported by small island states, but not by the US or the BASIC countries, and in the end, it became legally binding for parties to have an NDC, but not to fulfil it (Rajamani 2016). Ultimately, the PA again exemplifies the different perspectives of the EU and the BASIC countries. While the EU portrayed the Agreement as a major EU achievement (Arias Cañete 2016), few non-EU negotiators and experts see the EU as a central architect. Indeed, the perception is that the EU was a bit side-lined in Paris but swallowed its pride to avoid another failure like Copenhagen (Interviews 2018m, 2019b, 2019c, 2019d).

Many developing countries, and especially the BASIC countries, have difficulties with seeing the EU as a climate leader because it lacks widespread support in the developed world. From their perspective, the EU has not been able to establish a stable block of climate leaders within the group of developed countries. In the same way European negotiators often start to talk about China, the BASIC countries and the G77 when we ask about India or Brazil, BASIC country interviewees often start to talk about individual European countries, the US and 'the developed countries' when we ask about the EU (Interviews 2013c, 2014a, 2014e, 2017e, 2017h, 2019a). The varying climate action in other developed countries, like the US, Australia, Canada and Japan, has been a major disadvantage when the EU has tried to 'lead by example' (Interviews 2013a, 2014a). Even when the EU can show concrete reductions of own emissions and concrete cases of climate finance, the combined efforts of developed countries is inadequate compared to what is needed to attain the 1.5 or even two degree target. From this perspective, it becomes a pertinent question why the EU is pointing fingers at China and India when it cannot get its 'allies' in the developed world to reduce their luxury emissions (Shue 1993; Saran 2015). Like Copenhagen, Paris was a historical moment of climate-positivity in key countries. Yet unlike Copenhagen, the parties managed to reach a fairly common understanding of what would be achievable before the Paris negotiations started (Walker 2018; Interviews 2019b, 2019d). However, the crucial engagement of the US ended after Paris with the election of Donald Trump as President, leaving the EU as the only major developed party with serious mitigation ambitions. The question of EU climate leadership again climbed high on the agenda for European climate policy analysts. This being said, the whole climate leadership question is in essence a Western view on the climate negotiations. Being a climate leader is not a big issue in the BASIC countries' rhetoric.

A leader is something you are perceived to be, not something you claim to be. Although China clearly has leadership aspirations, it would not claim to be a climate leader or say that it has ambitions of influencing other parties' willingness to commit.

5.4 The climate regime after Paris: towards polycentricity

Notwithstanding the criticisms outlined above, developing country climate actors recognise the EU's seriousness in the negotiations, its willingness to commit to climate change mitigation and its recognition of equity as an important issue (Interviews 2013a, 2017g, 2017h, 2019a, 2019d). The EU member states have established a well-functioning system for themselves, based on legal structures with carrots and sticks. The EU climate policies function within this system with differentiated commitments for the attainment of a mutual goal. From the outside, the impression is that this way of organising European affairs has made European climate policy practitioners and experts too narrowly focussed on supranational legal frames as necessary for ambitious global climate governance (Interviews 2014a, 2019d). The PA's polycentric governance architecture with NDCs and inclusion of non-state actors' climate initiatives, punctuated the EU's ambition of scaling up the European system to the global level. However, although the developing countries do not perceive the EU to be a 'green leader', they do perceive the EU to be an essential and important party for global climate governance, especially after Paris. As one Brazil-based climate negotiations expert argues about the UNFCCC: 'Without the EU, the whole thing collapses. Developed countries have to be seen as leading in mitigation' (Interview 2019d).

The burden on the EU's shoulders is therefore large. To keep the multilateral negotiations alive and to ensure the engagement of large developing countries, the EU needs to keep ratcheting up its own ambitions, providing financial and other kinds of assistance to the least developed countries, and keep a good dialogue with major parties among both developed and developing countries. The developing countries' impression that the EU tried to dictate solutions for global climate governance in Kyoto and Copenhagen, stayed with the EU negotiators even after they embarked on a journey towards mutual recognition and non-domination (Interview 2019b). The Paris negotiations changed this impression, but the polycentric structure of the climate regime remains unfamiliar to the EU, despite its own experience as a multi-level polity.

One of the main disagreements between the developed and the developing countries after the PA follows in straight line from the disagreements at COP-8 in New Delhi in 2002: should the UNFCCC focus mainly on mitigation, or equally on mitigation, adaptation, loss and damage, finance and sustainable development? The EU and most developed countries support the mitigation focus, while most developing countries support the inclusion of various issues on the UNFCCC agenda. In its efforts to keep the negotiations on the mitigation-track and to eventually oblige large developing countries to mitigate, the EU is often perceived as a supporter of the US agenda (Williams 2019). Although the EU and the US disagree on several issues of procedural justice and when it comes to mitigation ambitions, they often agree on issues that the G77 regard as important in the negotiations. A key example is the difference between finance for adaptation and finance for loss and damage, which is regarded as non-important by the EU and the US. In the eyes of the developing countries, this disregard for the G77 position exemplifies postcolonial arrogance and seriously undermines the EU's potential for green leadership. Another recent example that illustrates why it is difficult for the developing countries to perceive the EU as a leader after Paris, is the debate on pre-2020 commitments. Pre-2020 refers to the commitments of developed countries before the PA's commitment period starts in 2020. At COP-23 in Bonn in 2017, the EU and other developed countries refused to have pre-2020 on the official agenda. This overriding of the COP agenda by developed countries created massive protests from developing countries and confirmed their longstanding impression that the developed countries are trying to dictate the agenda in their favour. While the EU argued that the negotiations needed to focus on the implementation of the PA, the G77 argued that it would not make sense to talk about post-2020 targets if the parties with pre-2020 targets had not attained those targets. While for some developing country parties the pre-2020 issue was probably mainly a possibility to stall progress in the negotiations, it is a real concern among developing countries that the developed countries are using their structural privilege in international politics to push the mitigation burden onto them to avoid own mitigation costs (Okereke and Coventry 2016, 14). In this picture, it is not possible for the EU negotiators to isolate their argument from the context they operate within. When an American, or a Swiss, or an Australian climate actor voices concern that more Chinese can afford to eat meat, or that more Indians can afford to own a car, or that more Latin-Americans can afford air travel, the EU is perceived to be part of that group; the developed countries that want to keep their privileged life style and 'luxury emissions' to themselves (Shue 2010).

5.5 Conclusion

To a great extent, perceptions of justice will depend on the perceiver's understanding of the world. From the perspective of developing countries, the main problem with the EU's role in climate justice is that the EU seems to assume that its understanding of the world is universal. Accustomed to being 'the others' in international politics, developing countries are used to their worldviews and normative preferences being questioned and often disregarded in global cooperation contexts. In the climate regime, the developing countries, and most vocally the BASIC countries, have resisted the developed countries' disregard of their political and comprehensive developmental understanding of the climate problem. As our interviews in the BASIC countries have shown, there is a large variety of views on how the BASIC countries perform in relation to climate responsibility and justice. Yet, on the question of the fundamental structural inequality in the current international order, the negotiators, the political opponents, the critical NGO activists, the energy bureaucrats, the researchers, and the energy businesspeople to a large degree close ranks. The processes of obtaining visas to participate in the COPs, the airport border controls, and the Western climate journalist's naming of G77 and BASIC as the UNFCCC 'troublemakers' are all incidents that remind the individuals and groups within and across developing countries of whose norms and rules of justice are allegedly universal.

Several scholars argue that although claimed to be universal, the traditional theoretical assumptions dominating IR analyses and also the climate justice literature are mainly developed in specifically western European and North American historical and cultural contexts, and thus are not necessarily fruitful for understanding the role of developing countries in international cooperation (Tickner 2003, 2013; Acharya and Buzan 2010; Steinberg 2012; Stuenkel 2016). One example is the liberal order, by the EU regarded as the best way to assure liberty and equal rights to people, but by others regarded as a way to impose, and make normatively unimpeachable, an international order that is beneficial to the former colonial powers. This claim has parallels with the spring 2020 protests under the slogan of 'black lives matter'. In choosing to remain unaware of their own historical privilege, the privileged, sometimes unknowingly and unintentionally, support norms that help maintain the unfair structural hierarchy.

This also pertains to our three justice conceptions, which may look different from a non-European perspective. The developing countries' experience of structural inequality in the international order begins with individuals and groups but is aggregated to the national level. Some of the distinctions between non-domination and mutual recognition thus become blurry. Although the BASIC countries' climate justice preferences are very state-centred and focus on sovereignty and non-interference, the fundamental recognition of difference and autonomy they call for is in line with mutual recognition. Mutual recognition's call for due hearing of all groups and individuals also goes for the aggregate, the respective states. The EU's focus on universal facts and numbers coming from the IPCC, has led to the ignorance of other facts; the bottom-up knowledge coming from the most vulnerable, explaining how climate change mitigation and adaptation is inseparably connected to fundamental socio-economic development in poor societies. This argument was first made by Indian Prime Minister Indira Gandhi in relation to environmental damages at the UN Conference on the Human Environment in Stockholm in 1972. Apart from the Swedish Prime Minster, Gandhi was the only state leader to attend the conference. Still, it was not until after the Copenhagen failure that the EU understood that the developing countries' perception of climate justice differs from the EU's, the lines between the individual and the state are blurrier, because the experience with injustice is both personal and structural.

Returning to the discussion of the EU as a normative power or supporter of climate justice, it is not necessarily wrong to argue that the EU has shown high ambitions and can claim the role of leader when it comes to implementing climate policies. But in claiming the power to define the climate problem, and offering adequate solutions to that problem, other understandings of the problem are effectively dismissed. Overcoming that breach with mutual recognition has been one of the main aims of developing countries in the climate negotiations. The trust and reciprocity between EU member states builds on a thorough processes of recognition of past atrocities within Europe (Diez 2005; Sierp 2020). But such a process has not occurred in relation to the colonial atrocities. Regarding impartiality and binding climate commitments, developing country negotiators do not think legally binding commitment will make any difference as to the unjust treatment of states in the international community. Even if there is no bindingness and no sanctions in the climate regime, they think they will be sanctioned somehow if they do not deliver what they promise in their NDCs. Similarly, they do not think developed countries will be sanctioned, even if they do not deliver on binding commitments (as the experience from the Kyoto Protocol has taught them).

At the same time, there is reason to worry about the BASIC countries' and other developing countries' ability to ensure climate justice to the most vulnerable within their own borders or in the LDCs and countries most affected by climate change. The BASIC countries have not adequately addressed domestic climate injustice, and their governments often dismiss the critique coming from domestic vulnerable groups and small developing countries. Receiving this critique from the EU, however, is hard to understand as a sincere attempt to change the current 'normal' in favour of the most vulnerable. The now so much debated statues of King Leopold II, Christopher Columbus, and Cecil Rhodes bear witness to the Europe that the EU is trying to forget (Hurst 2020). If the EU wants to have a role in shaping the new normal of climate justice, a good first step would be to acknowledge its role in shaping the current international order, perceived to be structurally unjust by many climate actors in developing countries.

Note

1 Expression used by a delegate from a small African country in a hallway conversation during COP23 in Bonn in 2017.

6 Solidarisation

The productive ambiguity in the EU's climate policies

> They said Europe is too complicated to agree quickly. They said we had too many hoops to jump through. They said we were all talk. They even started to question whether our heart was really in it. Today we clearly showed that we mean business. [...] Today's agreement shows unity and solidarity as Member States take a European approach, just as we did in Paris. This is what Europe is all about. In difficult times, we get our act together, and we make the difference. [...] We are reaching a critical period for decisive climate action.
>
> (European Commission 2016c)

This comment by the Climate Action and Energy Commissioner, Miguel Arias Cañete, on the European Union (EU)'s ratification of the Paris Agreement (PA) illustrates that the EU's role in the global climate change regime is difficult to assess. As *Chapters 3* and *4* exemplify, the EU itself makes strong claims about its self-understanding as a strong vanguard for effective climate action. And yet the statement's defensive character reveals that this positive self-assessment is contested – as *Chapter 5*, particularly, has shown – by other political actors as well as within the academic community. Can the EU be considered a green normative power, and does it as such promote climate justice in international society? Or does it fail to do so because of the many inconsistencies and insufficiencies that have provoked substantial criticism, which we have also raised and addressed throughout this book?

The aim of this chapter is to discuss this question by combining our conceptual framework of climate justice with the debate about the EU's contribution to the so-called *solidarisation* of international society. Within the international society approach of international relations, solidarisation denotes a process of change, which helps to transform international order towards less state-centric structures away from what is called a pluralist international society (Buzan 2004; Ahrens and Diez 2015; Ahrens 2019). In the context of climate change, such a transformation is paramount because climate change undermines the logic of a state-centric order. At the same time, however, the combat against climate change necessitates state action. It thus requires and contributes to the reproduction of state-centric structures.

However critical one may be of the oftentimes thwarting role that states and their particular national interests play in global climate action, the realities of the current international order are such that an effective climate regime can only work with and through states, rather than fully against them. This is a fundamental ambiguity that poses intriguing tensions to climate justice considerations, because both dimensions of order – a solidarist one and a pluralist/ state-centric one – entail their own legitimate justice claims. It also mirrors the productive tension between our three conceptions of justice and underlines the claim that a reductionist focus on either non-domination, impartiality, or mutual recognition is problematic (see *Chapter 2*). Combining our three conceptions of justice with the pluralism-solidarism debate therefore provides an important supplement to a comprehensive analysis of the EU's contributions to the climate regime and its normative and justice-related implications. The main argument is that the EU's strategic move away from mere impartiality towards increasingly including claims of justice as non-domination and mutual recognition seems to suggest that the EU has backpedalled on its ambitious solidarist agenda in the climate regime. However, as the discussion of the link between justice and solidarisation indicates, this conclusion falls short of recognising the fundamental underlying ambiguity that the pluralism-solidarism debate as well as the three conceptions of justice pose to the issue of global climate change. Putting forward this argument, the following caveat is important: it is in no way the intention of this chapter to whitewash the EU's involvement in the global climate change regime or reason away some of its obvious failures. Yet the argument is that it is crucial to understand the unresolvable ambiguities that justice claims as much as a solidarist agenda inevitably entail, and to recognise the possibilities and constraints they imply for the EU's engagement with the global climate regime.

The chapter proceeds as follows: we will first elaborate on the concept of solidarisation and explain in more detail the relevance of the pluralist-solidarist debate for climate change as an issue of global governance (Section 6.1). Section 6.2 then brings together our three conceptions of justice with solidarisation and looks at the evolvement of the global climate change regime and the EU's role within it by combining both perspectives. Rather than repeating the empirical analysis, this section draws on the findings outlined in *Chapter 3*, namely the EU's shift from pursuing mainly impartiality at first to including stronger claims of non-domination and mutual recognition. Here, we illustrate that solidarisation – while in principle desirable – does not consistently equal the enhancement of justice. Instead, some solidarising moves promote one form of justice and run counter to another at the same time. Likewise, while justice claims in all three justice conceptions are to some extent legitimate, some of them tend to support solidarisation, while others reproduce a problematic state-centric order. Subsequently, Section 6.3 sets out the implications of the solidarisation-justice nexus and of the role of ambiguity within it. The section thus sums up how both perspectives provide an alternative understanding of the EU's learning process.

6.1 Solidarisation and climate change

States form a *pluralist* international society if their cooperation rests on a thin basis of shared norms and values. Maintaining stability within the largely state-centric order, which protects difference and diversity among its members, is the predominant purpose. In this conception, states are the primary actors and their national sovereignty is the most fundamental principal of order. In contrast, a much thicker basis of shared norms and values underpins a *solidarist* international society. Here, universal values and ideas, which are promoted and protected beyond state borders are constitutive. As a consequence, solidarism challenges the role of states as primary actors and as sole bearers of rights and duties. Individuals and other non-state-actors become subjects to international law. Their rights require protection – if necessary, against the claims of sovereignty of a particular state. Solidarism thus ultimately means to transfer classical features of domestic societies to the international level (Bull 1966, 2002, 230–32).

The pluralist-solidarist debate entails an analytical as well as a normative dimension. As analytical categories, they help to study a given order at a particular point in time and to examine whether change in one or the other direction has been occurring. Yet, more often than not, they are used to express a normative preference for a particular order. For the analytical as much as for the normative dimension, it is crucial to note that pluralism and solidarism are not fully distinct categories, but they are ideal types at the opposite ends of a continuum (Weinert 2011). Thus, while a certain order can be more or less solidarist or pluralist, it will never be purely one or the other.

Pluralists argue that international society cannot be based on more than a rather thin basis of shared norms and values. Lacking large-scale solidarity among states, such a basic order is all that is feasible and therefore desirable. Solidarists in contrast try to promote a thicker basis of shared norms, including individuals' rights. The normative desirability of such a position seems compelling. A classic example is the Responsibility to Protect (R2P), which challenges the sovereignty of states if they fail to protect their citizens against human rights violations and at the same time legitimises other states and institutions to intervene, if necessary. Yet pluralists reject this argument on the ground that attempts to enlarge the shared (minimalist) normative basis of international society puts the overall order into jeopardy. Some argue, therefore, that a pluralist conception is indeed not only the more realistic one, but also the one that is ethically more desirable (Williams 2005). Indeed, our discussion of the disputes between the EU and developing countries in *Chapter 5* has demonstrated the normative value of such a position and highlighted the problems of a solidarist approach, which at the same time seems necessary to achieve real progress towards climate change mitigation.

There is thus no easy way out of the pluralism-solidarism debate. Both positions rest on valuable and valid arguments. And while it still makes sense to lean towards one or the other pole when debating particular issues of world politics, both positions will be constantly present – and inevitably so. This inevitability of the parallel existence of pluralist and solidarist structures is what we call a fundamental ambiguity of international society (Ahrens 2018b).

This ambiguity is particularly important in the context of climate change, as the following quote by Hedley Bull summarises precisely:

> In the long run it is unlikely that action at the purely state level will be sufficient to cope with environmental dangers, and the functionality of the states system [...] will depend upon the emergence of a greater sense of human cohesion than now exists. However, the idea that the states system should be regarded as obstacle to the development of this greater sense of cohesion, rather than as the means through which it may come to take shape, is an unhelpful one.
>
> (Bull 2002, 284)

Bull's assessment illustrates two things. First, even a basic pluralist international society will need to provide some instruments to adapt to and mitigate climate change because its major goal is the maintenance of a state-based order and the survival of the sovereign members of this very order. In the form of rising sea levels, the increase in severe natural disasters and resulting flows of migration, global warming posits a massive challenge to such a purely state-based international order. Thus, even without invoking any individuals' rights, a pluralist international society should have a self-interest in accommodating some mechanisms to combat climate change. And yet, although leaning towards the pluralist pole himself, Bull fully acknowledges that some moves towards solidarism will be paramount to maintain the very functionality of the state system as such.

Andrew Hurrell positions himself much more clearly towards solidarism:

> Whatever view one takes about the state itself, the ecological challenge has undoubtedly served to call into question both the practical viability and the moral acceptability of a state-based pluralist international order.
>
> (Hurrell 2007, 222)

There are several reasons for the indispensability of solidarisation in an effective climate change regime: the problem of climate change per se is non-territorial in character and thus sits uneasily within a state-centric order. Neither the detrimental effects of Greenhouse Gas (GHG) emissions nor the benefits of GHG emission reduction follow any territorial logic – although of course the negative effects and benefits, as discussed in *Chapter 2*, vary decidedly between states due to their different vulnerability and coping capacity, which again is highly dependent on the existing international order. Furthermore, the norm of environmental protection is rarely limited to a state-centred logic in the political discourse, but instead tends to invoke the rights of individuals, present, and future. The link between climate change and the fulfilment of individual human rights is increasingly acknowledged on the international agenda (OHCHR 2015).

The second aspect that Bull's statement illustrates is that – the indispensability of solidarisation notwithstanding – the current international order faces pluralist realities. These must be considered when promoting legitimate moves to further solidarise international society in the context of climate change. This was certainly true at the time of Bull's writing at the end of the 1970s, but it has not lost its relevance today despite major developments in the climate regime. On the contrary, it seems that the liberal era of the 1990s and early 2000s, in which various forms of trans- and supranationalism have become more prominent within international society, has come to its end. Today, we witness a renaissance of state-centrism. The national level is back at centre stage. This has put the EU in a difficult position to adhere to its supranational ideas of a top-down regulated climate change regime as it had favoured particularly up until the failure of the Copenhagen summit in 2009. Under these conditions, it is all the more difficult to effectively enhance the climate regime, while not taking pluralist, state-centric concerns into account.

As the following section will demonstrate, a normative assessment of this situation is much more problematic than it appears at first glance. Invoking justice claims of individuals who suffer from the consequences of global warming or whose survival is put at risk, pushes a strong and straightforward solidarist agenda that promotes a regime in which 'the development, the well-being, and the dignity of the individual human being are a matter of direct concern to international law' (Lauterpacht 1946, 27; Sachs 2014, 220). However, given the normative ground of pluralist structures, a variety of other justice claims comes into the picture. Our three conceptions of justice help to address the seemingly contradictory justice implications of solidarisation, as evidenced in *Chapter 5*, in a more systematic manner.

6.2 Climate justice and the normative ambiguities of solidarisation

6.2.1 Identifying solidarisation

To spell out in detail what solidarisation means in the context of climate change, we set forth three indicators of solidarisation, which we extract from the broader English School literature. To systematise the empirical analysis, we focus on solidarisation in two different spheres, i.e. international law and diplomatic practices. Both are fundamental constituents of international order (Bull 1977; Buzan 2004; Holsti 2004).

First, *extension of rules of cooperation*: in his conception of international society, Hedley Bull suggests a tripartite typology of rules, which are constitutive to international order (Bull 2002, 64–86). The first two layers of rules are 'constitutional normative principle[s] of world politics' and 'rules of coexistence'. These types of rules define the basic ordering principle of international society and specify some very basic standards to ensure that a society can exist at all. These include the limitation of violence, some property rights, and a certain degree of reliability of agreements. Bull's third level of rules are those that are directed towards more complex, far-reaching, and advanced goals of an international society. This is where solidarisation potentially sets in (Buzan 2006, 83). Whenever there is the endeavour to move beyond mere rules of coexistence and to push for the development of more far-reaching regulations, then this is evidence for a solidarising move and an indication of solidarisation, whenever this move is at least partly successful.

In the climate regime, this indicator embodies a rather light version of solidarisation. It is less fundamental than the other two because it neither implies the challenging of states as primary and exclusive constituents of international society nor a re-interpretation of sovereignty as the most elementary principle of international society. Regarding international law, this indicator would mean an enhancement of the cooperation to strengthen the regime. Any push for the development of more far-reaching regulations to bolster the combat against global warming constitutes solidarisation. The EU would contribute to such processes if it acted as a leader in the negotiations that pursues ambitious targets to be commonly agreed upon. Regarding diplomatic practices, the EU would contribute to solidarisation if it worked towards increasing and diversifying diplomatic contacts and intensifying the overall diplomatic activity.

Second, *enhanced role of non-state actors*: two aspects are important for this indicator. First, solidarisation occurs if non-state actors play an increasingly important role in international politics and thus challenge the primacy of the state (Buzan 2004, 48). For instance, the increased relevance of Non-Governmental Organisations (NGOs) in international politics and the enhancement of their substantive contribution to outcomes of multilateral negotiations entail solidarisation. Second, and more specifically, there is evidence for solidarisation if the significance of individuals vis-à-vis the state is increased and if thus individuals themselves become subjects of international law. The more international legitimacy is based on the rights of individuals rather than primarily on the rights of states (Armstrong 2006, 132), the more an international society becomes solidarist. The status of states as primary bearers of rights and duties and their role as most crucial entities in international society is mitigated (Vincent 1986; Wheeler 1992, 468; Wheeler and Dunne 1996, 95; Linklater and Suganami 2006, 64; Bain 2014, 160).

In the climate change regime, solidarisation in this sense would mean to challenge or even partly overcome the thoroughly state-centric logic of the regime and its underlying international law provisions. For instance, solidarisation is enhanced through non-state actors and individuals becoming a reference point for the international debate about global warming. Establishing a clear link between climate change and human rights likewise constitutes a solidarist move. Furthermore, the active involvement of non-state actors in the diplomatic process of climate negotiations entails solidarisation. In contrast to the first indicator, this one captures such involvement of non-state actors that go as far as to challenge the primacy of state actors in the diplomatic process.

Third, *re-interpretation of national sovereignty*: in a more solidarist international society, the meaning of sovereignty as a constitutive principle changes fundamentally. National sovereign rights are only indefeasible to the extent that they are tied to states' responsibility for their own people (Vincent 1986; Wheeler 2000; Reus-Smit 2001). Sovereignty becomes a conditional rather than an absolute principle (Vincent 1986, 113; Buzan 2014, 124). With the establishment of the International Criminal Court (ICC) or the adoption of the R2P, the world has witnessed such shifts of the meaning of sovereignty being inscribed into international law provisions.

Following this indicator, the climate change regime would become more solidarist if there were more stipulations in international law that decrease the opportunities of states to use national sovereignty concerns as excuse for non-action. Furthermore, solidarisation would mean to work towards the implementation of international law principles that emphasise commonalities between states and similarities of how they are affected by global warming rather than constantly pointing to differences among them. With respect to the diplomatic process, solidarisation occurs through the establishment of institutional mechanisms that aim at exposing states' sovereign action regarding climate change to the scrutiny of the international community.

The following *Table 6.1* summarises the three indicators of solidarising processes in the climate regime.

As these indicators clearly show, solidarisation above all implies a move toward greater impartiality. The strengthening of international law, the enhancement of the rights and duties of individuals, and the reformulation of sovereignty all make states subject to a higher level of governance that takes decision based on presumed objective necessity rather than negotiations between states and overrides their claim to non-domination. Yet, especially the changes to diplomatic processes also open solidarisation towards a strengthening of mutual recognition in that they strengthen the position of diverse civil society actors.

Table 6.1 Solidarisation in Climate Change

	Solidarisation in…	
	… international law provisions	*… the diplomatic process*
Extension of rules of cooperation	– Promoting cooperation or safeguarding its continuation – Ambitious regulations and clear emission reduction targets – Working towards strengthening the regime through reaching agreement on clear regulations and universal treaties – Promoting the precautionary principle in order to move cooperation forward	– Intensification of diplomatic contacts and outreach to all relevant stakeholders
Strengthened Role of Non-State Actors	– Challenging of a purely state-centric logic of the regime – Promoting individuals as reference point – Promoting the precautionary principle in international law as an obligation to act for the well-being of people – Highlighting the link between climate change and the protection of individual human rights	– Fostering the active involvement of non-state actors in the diplomatic process – Active outreach towards civil society and non-state actors – Effective coalition-building with civil society in climate change negotiations
Reinterpretation of sovereignty	– Promoting the precautionary principle in order to prevent national sovereignty concerns being used as excuse for non-action – Overcoming a regime logic that emphasises differences between sovereign states rather than commonalities in how they are affected by global warming	– Working towards the establishment of institutionalised diplomatic practices that expose states' sovereignty by putting their action within the regime under scrutiny of the international community

Adapted from Ahrens (2018b, 155).

6.3 Solidarisation and justice in the climate change regime

6.3.1 The EU's role in the general advancement of the climate regime

From the perspective of solidarisation, the 'advancement of the climate change regime' can mean two different, sometimes opposing things. On the one hand, solidarisation can be put into practice by agreeing on binding rules that are enforced by global multilateral institutions. As *Chapter 3* has demonstrated, the EU, up to Copenhagen, had aimed for such rules but faced opposition. Thus, the United Nations Framework Convention on Climate Change (UNFCCC) and the Kyoto Protocol contained limited levels of solidarisation in substance and scope. On the other hand, 'advancing the regime' in a more modest sense is about reaching a new universal agreement at all. Even if such an agreement does not, for instance, impose clearly defined reduction targets for all states, it can still enhance the degree of cooperation by setting out pathways and mechanisms for such cooperation in the future, and thus moves the climate regime further, although perhaps only marginally, towards the solidarist pole.

These two meanings correspond to the *depth* and *scope* of the solidarising process, which entails a trade-off between these two dimensions (Voigt 2016, 161). Up until the 15th Conference of the Parties to the UNFCCC (COP) in 2009, the EU took a very principled stance on both dimensions. As *Chapter 3* showed, this was not successful. The Copenhagen Accord hardly entails any solidarisation, neither in terms of its scope, nor in relation to the depth of its regulatory power. In the aftermath of Copenhagen and in the run-up to Paris, the EU compromised on the depth dimension by being more flexible in its position on a top-down approach (Oberthür 2016, 4). This, however, did not weaken its solidarist agenda on the scope dimension to come to a universal and binding agreement that would be endorsed by all parties. The continued high share of argumentations linked to justice as impartiality in *Figure 3.1* underlines this conclusion.

A major cornerstone of the EU's position in international negotiations has always been to promote multilateralism within the climate change regime in order to reach a universal agreement at the United Nations (UN) level 2 (Schunz 2012, 2015). Such an approach has been put under jeopardy with the growing bifurcation of the regime, which we discussed in *Chapter 5*. There were points in time when a splitting up of different strands of cooperation in the climate change talks seemed likely. The Bali Action Plan (UNFCCC 2007), the outcome document of the 2007 climate summit, established two different negotiation tracks under the auspices of two Ad hoc working groups – one on 'Long-term Cooperative Action' (AWG-LCA) and one on 'Further Commitments for Annex I Parties under the Kyoto Protocol' (AWG-KP).[1] Both working groups focussed on different parts of the negotiations: whereas the AWG-KP mainly focussed on reduction commitments of industrialised countries under a second period of the Kyoto Protocol, the AWG-LCA was concerned with issues, such as commitments for emerging economies, obligations of industrialised countries to support adaptation measures as well as technology transfer and cooperation from industrialised to developed countries (Bals 2009, 7; Jordan and Rayner 2010, 72; Schunz 2012, 17). It remained unsettled in Bali whether these two tracks should co-exist for the time being, but then be merged in one comprehensive legally binding agreement to be reached in Copenhagen, or whether they should result in two differentiated agreements (Bäckstrand and Elgström 2013, 1377). Thus, it was by no means self-evident that the climate talks would lead to a single comprehensive treaty at some point, which was the clear EU preference (Oberthür 2011, 669).

Two statements illustrate the conflictive positions in the negotiations. A statement on behalf of the Group of 77 (G77) and China at an AWG-LCA session in 2008 expounds that '[t]he AWG KP and the AWG LCA are two complimentary parallel processes' (G77/China 2008). The statement furthermore emphasises that both AWGs are based on a *different* mandate and pursue *different* tasks. It thus voiced the reluctance of developing countries to pursue 'a single outcome or one-track approach as it would undermine the dichotomy between developed and developing countries' (Bäckstrand and Elgström 2013, 1377). In contrast, the EU has insisted on merging both negotiation tracks in order to reach a common and universal agreement:

> We are committed to reaching a global, ambitious and comprehensive agreement here in Copenhagen. An agreement which keeps global warming below 2 degrees and which encompasses all nations of the world. It should be detailed, cover all the Bali building blocks, enable immediate action and build the best possible foundation for a long term cooperative effort.
>
> (EU Council 2009, 15)

The EU made clear that it 'can not see a scenario coming out of Copenhagen where there is a new binding agreement solely in the Kyoto Protocol' (EU Council 2009, 14), meaning that eventually only the AWG-KP negotiation track would lead to an agreement with new commitments being confined to industrialised countries.

Thus, the establishment of two different negotiation tracks, endangered the scope dimension of the solidarising process because it impinged upon the universality of cooperation by furthering the bifurcation of the regime. Eventually, the EU has played a crucial role in bringing about a major change in Durban (COP-17) in 2011 (Oberthür 2016, 3–4). The *Durban Platform for Enhanced Action* set out to end the Bali negotiation tracks by 2012 and initiated a new negotiation process to be finalised in Paris 2015. This new process 'makes no obvious distinction between developed and developing nations' (Hurrell and Sengupta 2012b, 472) and instead demands 'the widest possible cooperation by all countries' (Durban Platform, quoted in Hurrell and Sengupta 2012b, 472). This formulation illustrates well that the overcoming of the regime's dichotomisation entails solidarisation because it enhances the level of cooperation. For the EU, agreeing in Bali to the establishment of two negotiation tracks, was a necessary concession that enabled the continuation of the negotiations at all. Meanwhile, the EU had never abandoned the aim of overcoming this split and constantly promoted the goal of reaching a comprehensive agreement, universal in scope.

In *Chapter 3*, we have interpreted the EU's deviation from its principled preference for a binding regime as a shift from impartiality towards more non-domination and mutual recognition. Yet we also demonstrated that this shift was never complete, that articulations of impartiality still outnumbered those of the other two justice dimensions, and that this ultimately made Paris possible. This is corroborated by our analysis of solidarisation as enhancing the climate regime. It was the seemingly contradictory pursuit of both a globally binding regime and the increasing respect for other perspectives that ultimately made progress possible – although of course not achieving a maximalist impartial goal. If the EU had insisted on its regulatory approach, this would have been evidence of a strong solidarist agenda. However, this also would have put solidarisation in jeopardy altogether because, as *Chapter 5* has shown in particular, large developing countries would not have accepted binding top-down regulations especially if adopted and enforced within the current international order, which they perceive as inherently unjust. An ambiguous concession, coupled with a strategic weakening of securitising moves (see *Chapter 4*), thus made solidarisation possible.

6.3.2 CBDR: increasing distributional justice but decreasing effectiveness

In *Chapter 3*, we have argued that the principle of Common But Differentiated Responsibilities (CBDR) resembles a weak version of mutual recognition, while it also entails elements of non-domination. It speaks to the fact that the responsibilities as well as the capabilities of different state actors to combat climate change are unevenly distributed. From a justice perspective, thus, there are good reasons to defend the CBDR because it allows promoting the idea of distributional justice in the climate regime. *Chapter 3* also states that the EU has contributed to implement the CBDR, for instance in the 1992 Rio Declaration.

Complementing the justice analysis of the CBDR with a solidarisation perspective is revealing because it further brings to the fore the underlying ambiguities of this principle. Furthermore, it demonstrates that the EU took an apparently ambiguous stance on the CBDR, which ultimately may have contributed more to the advancement of the climate change regime than a clear-cut defence or rejection of the CBDR would have.

The CBDR, while in line with the idea of distributional justice as well as mutual recognition and non-domination, poses severe problems to the regime's effectiveness. To accommodate the CBDR, the UNFCCC and later the Kyoto Protocol have introduced different annexes that differentiate countries in disparate groups: developing countries are supposed to focus on adaptation and are entitled to appropriate financial and technical support due to their restricted capacities. Developed countries, in contrast, shall concentrate on mitigation measures and thus on actual emission reduction. While differentiation must certainly be a legitimate concern in international climate negotiations, this clear-cut classification of different kinds of countries is problematic. It reproduces an international society marked by severe differences between different groups of states, rather than taking into full consideration the fact that global warming is a problem that affects the international community as a whole. In this sense, the concrete realisation of the CBDR principle in the climate change regime reflects a pluralist, state-centric international society.

Yet such a state-centric interpretation of the CBDR is not a logical necessity. One could also imagine differentiated responsibilities of individuals (see e.g. Neumayer 2000; Caney 2005, 756) or non-state actors (Bulkeley and Schroeder 2012; Bäckstrand et al. 2017), which would connect the CBDR to impartiality or strong mutual recognition. Despite these theoretical considerations, however, the dominant practice of the regime clearly links the CBDR to different states. The resulting bifurcation within the climate change regime has prevented further progress and change towards effective mitigation of global warming and has ultimately been a major source for the deadlock of international climate talks. As a commentator of the European Centre for Development Policy Management has pointed out shortly before the Paris summit: 'CBDR ... could be a huge barrier to a strong settlement' (ECDPM 2014).

While the EU has actively contributed to the incorporation of the CBDR in the Rio Declaration, this happened reluctantly and from a fairly sceptical stance towards this principle, exactly because of its links to non-domination (Bäckstrand and Elgström 2013, 1675). The EU supports the idea that the climate change regime must accommodate different responsibilities and capacities (Interviews 2017a, 2017b; 2017d). Yet it rejects the rigid dichotomy between groups of states that had been put in place and that had been further reinforced by the Kyoto Protocol. The EU, therefore tried to repel the dominant pluralist form of differentiation as manifest in the CBDR. An important partial success was reached at the Durban summit (COP-17) in 2011, as the Durban Platform for Enhanced Action makes no reference to the CBDR as a core principle of the regime (Hurrell and Sengupta 2012b, 427).

The CBDR also was a major source of contestation in the negotiations that led to the adoption of the PA. While the PA does mention the CBDR in its Preamble (Articles 2 and 4) it 'makes no fundamental distinction between developed and developing countries and provides that all Parties shall account for their contributions' (Obergassel et al. 2016, 35). Most notably, while acknowledging differentiation as a legitimate and necessary part of an international climate deal, the Agreement does not mention the Annex-I and non-Annex-I categories and therefore manages to incorporate developing states' concern for historical responsibilities of industrialised countries, while at the same time not exempting developing states from reduction commitments. In this sense, the PA has managed to pay attention to 'dynamic aspects of differentiation' and 'has struck a careful balance between the need for ambitious climate action and for fair effort sharing among parties based on differentiation' (Voigt and Ferreira 2016, 285).

Moreover, this may be considered as a first step to disassociate the CBDR from its ties to states and a pluralist order towards strengthening a differentiation between different groups and thus enhance its contribution to strong mutual recognition and impartiality. Contributing to this outcome, the EU rejected 'unacceptable bifurcated proposals [by the G77] for quantified commitments for public finance by developed countries only' (internal delegation document quoted in Dimitrov 2016, 5), although it was less hard-line in taking this position than the US, which was more focussed on safeguarding its own position in the regime (Dimitrov 2016, 5). This further corroborates our claim that an ambiguous, pragmatic attitude is more effective than an unreflected pursuit of impartiality. As Oberthür (2016, 4) notes, the EU 'taking a moderately progressive stance' was crucial in reaching both the Durban and the Paris outcome.

6.3.3 *Precautionary principle: strengthening impartiality and pre-emptive reasoning*

'When human activities may lead to morally unacceptable harm that is scientifically plausible but uncertain, actions shall be taken to avoid or diminish harm' (UNESCO 2005, 14). In the context of climate change, this definition of the precautionary principle means that there is an obligation to take effective measures against global warming if there is the risk – even a small one – that global warming might cause harm to the environment or to human beings (Bourguignon 2016). Thus, the principle prevents that the lack of full scientific evidence for the harmfulness of global warming can be used as a legitimising reason for non-action (Gardiner 2004a, 564).

In *Chapter 3*, we have linked the precautionary principle to justice as impartiality because it establishes a universal standard, emphasises the importance of scientific advice and international law and considers the rights of individuals. The EU is a strong promoter of precaution and it is stipulated in Article 191 of the Treaty on the Functioning of the European Union (TFEU) as a core principle of European environmental policy. Based on its role in European Court of Justice (ECJ) case law, Douma (2000, 141) even concludes that within the EU 'it is preferable to describe [the precautionary principle] [...] as a legally binding norm' rather than simply a general policy guideline. Despite the importance of precaution as a principle in EU policy, the PA does not make any explicit reference to the principle, which indeed suggests that the EU has stepped back from its strong focus on impartiality in the Paris negotiations. Yet the solidarisation perspective adds another layer to the analysis and clarifies the role of ambiguity in this context.

The precautionary principle is a manifestation of solidarisation according to all three indicators. The precautionary principle calls for further cooperative action among states even in case of scientific uncertainty. It follows therefore not only a preventive logic (averting an imminent danger), but a pre-emptive one (averting a potential danger, even in light of uncertainty whether this danger will materialise), underpinned by securitisating moves as analysed in *Chapter 4*. The second indicator, too, applies here, to the extent that invocations of the precautionary principle refer to individuals' well-being and the necessity to protect those individuals from harmful effects. Finally, the precautionary principle establishes a close link between sovereignty and responsibility. If the precautionary principle takes effect in the practice of international law, it creates an obligation for states to act, even if such action conflicts with their immediate interests. Both, its pre-emptive logic as well as its implications for sovereignty make precaution a 'quite radical principle' (Hurrell 2007, 225).

Given the level of solidarisation that the precautionary principle entails, it does not come as a surprise that the principle is contested on the international level. Nonetheless, the EU has been successful in including the principle on various occasions in international negotiations. For instance, the EU was the leading force behind the inclusion of precaution in the Cartagena Protocol to the Convention on Biodiversity in 2000 (Vogler 2005). Furthermore, during the Rio summit in 1992, the European Community (EC) delegation suggested and promoted the precautionary principle most actively. Following the EC's suggested formulation, it was finally included as principle 15 in the Rio Declaration (Vogler 2005, 843) and was also adopted as Article 3 (3) of the UNFCCC.

In line with the findings of *Chapter 3* about a decrease in the support for impartiality in the PA, we could argue that the omission of the precautionary principle in the agreement indicates that the EU was less successful – or even less ambitious – in promoting its solidarist agenda. However, this conclusion is premature. The Paris process as a whole reflects the idea of precautionary action as its purpose is to set forth institutional mechanisms and pathways to respond to the uncertainties of an unhalted rise in global temperature. As some observers have pointed out (Morales 2015), the precautionary principle is implicitly entailed in the PA through its ambition mechanism. We will discuss this further in the section on the PA below. Here we can put on record further evidence that ambiguity has played an important role in the climate negotiations: not including the precautionary principle in the legal text of the PA may have helped to adhere to it – and thus to a degree of impartiality as well as to solidarisation – rather than proving that it has been abandoned.

6.3.4 Diplomatic strategies and practices

In the previous sections, the relation between justice and solidarisation was mainly such that an increase in solidarisation more or less equalled an increase in justice as impartiality. In this section, we demonstrate that this is not the necessary relation between the two perspectives. This poses an interesting tension between our conceptions of justice and solidarisation, which once again testifies to the significant role of ambiguity in this context, and the need to balance the three justice principles rather than play them off against each other.

Chapter 3 has revealed a remarkable change in how the EU presented itself as a diplomatic actor in the climate negotiations with the failure of COP-15 marking a major turning point. With a change in its diplomatic strategy in the run-up phase to Paris, 'the EU became a grown-up', as a diplomat from the European External Action Service (EEAS) pointed out (Interview 2017c). This shift in the EU's diplomatic strategies is a manifestation of an increase of mutual recognition and to some extent also non-domination. However, the change of diplomatic practices can at the same time be read as a major process of solidarisation.

Prior to the Copenhagen summit the EU's diplomatic strategy relied mostly on the normative persuasiveness of its own position and its 'leadership by example'. This has led to a lack in active coalition-building and outreach campaigns as well as to a disregard of different pathways to abate climate change. This has changed after COP-15: active outreach to and through non-state actors was essential in the EU's realignment of diplomatic activity. It was the EU's declared goal to enhance elements of public diplomacy, i.e. outreach to foreign publics (EEAS and European Commission 2013; Davis Cross 2017, 10, 16). In this process, the EU also promoted the role of celebrities in gaining public support for a strong and far-reaching new international climate deal (Neslen 2015). Prepared by the EEAS and its Green Diplomacy Network (GDN) as well as the Directorate-General for Climate Action (DG CLIMA), the EU implemented the *EU Climate Diplomacy Action Plan* of early 2015, which addresses outreach not only to governments but to the wider public as an important strategy (European Commission 2016b; Oberthür and Groen 2017, 13). As part of its diplomatic activity for climate change, the EU had also established the *Global Climate Change Alliance (GCCA)* already in 2007 (for a critical discussion see von Lucke 2020a). In 2014, a new phase of GCCA was launched under the *GCCA+ flagship initiative*. This updated version of the GCCA testifies to the solidarisation of the EU's diplomatic activity by explicitly promoting '[e]nhanced cooperation with Non–State Actors and Civil Society Organisations as well as new alliances with new stakeholders such as the private sector' (GCCA 2015a, 3) as a new feature.

Furthermore, similar to the EU's ambiguous stance on the CBDR, its realignment of climate diplomacy has contributed to minimise the regime's bifurcation between developed and developing states and has aimed at strengthening the impartial components. Much more than prior to Copenhagen, the EU in the run-up to Paris, organised active outreach and close exchange to third states, in particular a number of developing and most vulnerable countries, such as Small Island Development States (SIDSs) and Least Developed Countries (LDCs), through various diplomatic channels and for a (see *Chapters 3* and *5*).

For instance, the GCCA/GCCA+ aims at strengthening dialogue and co-operation with developing countries by providing an exchange structure, and also financial support to developing countries. Through a number of programmes in more than 30 countries, the GCCA intends to mainstream climate change considerations in development processes at an early stage (Davis Cross 2017, 12). This contributes to intensifying diplomatic cooperation and to overcoming major differences, and thus helps fostering solidarisation.

Furthermore, the GDN, which was relaunched prior to the Paris negotiations, is crucial in this context and another significant manifestation of solidarisation (Torney 2015b). Through local branches in third countries, the GDN aims at initiating outreach campaigns and consultations in these countries in order to gather information about other stakeholder's positions on climate issues as well as to promote the EU's position. This improved dialogue with third countries has enhanced the EU's impact on the negotiations (Davis Cross 2017, 11) and played a crucial role in establishing the Durban Platform – a first significant step towards overcoming the regime's dichotomisation (Obergassel et al. 2016, 35; Oberthür 2016, 3–4; Davis Cross 2017, 11).

Another crucial network for bridge-building between developed and developing states was the *Cartagena Dialogue for Progressive Action* (Biedenkopf 2016; Oberthür 2016). The same informal structures and diplomatic links culminated during the Paris summit in the formation of the *High Ambition Coalition (HAC)*, which played a major role in bringing about the PA (Interviews 2017b, 2017d, Oberthür and Groen 2017). Together with Norway, the EU played a crucial role in forming this group of ambitious countries from different regions (Brun 2016, 121) – a process that was already being prepared since the Durban summit in 2011. Initially an alliance between the EU and a number of smaller developing countries, the coalition managed to gain support by the US, Japan, and Brazil, the latter being particularly important because it split up the traditionally non-progressive BASIC group (consisting of Brazil, South Africa, India, and China) (Obergassel et al. 2016, 10). As an EU official reported, 'even China was on board [of the HAC] – even if not formally' (Interview 2017b). To take 'the lead in bridging between developed and developing states' (Interview 2017b) was the EU's major objective behind such activities.

In summary, the EU's change in diplomatic practices has enhanced the relevance of mutual recognition and non-domination. On the one hand, this has weakened justice as impartiality because hearing the voices of many actors, recognising different ideas, and ultimately also accepting the red lines of other states when it comes to obligatory commitments makes it more difficult to promote universal solutions and principles. On the other hand, however, the same process has entailed solidarist developments within the regime. And this has ultimately been a crucial step to bring about the PA.

6.3.5 The Paris Agreement: introducing impartiality through the pluralist backdoor

Much of what we have discussed in the previous paragraphs crystallises in the debate about the PA. Is the latest climate treaty ultimately the breakthrough that the advocates of a strong climate regime had hoped for, is it a step forward at least? Or, is it basically a paper tiger that allows states to pretend action, when they actually stick to business as usual? The combination of our justice conceptions and the solidarisation approaches provide an interesting perspective on this question.

As we have outlined in *Chapter 3*, the PA entails a weakening of justice as impartiality. Likewise, Robert Falkner has a point when he argues that the PA 'signals a retrenchment of solidarist ambition and a reassertion of a pluralist logic of decentralised coordination that protects national sovereignty' (Falkner 2018, 40). However, this does not tell the whole story. The PA does incorporate pluralist reservations against far-reaching solidarist change and it is not upfront with the promotion of an impartial conception of justice. However, the shift towards mutual recognition and non-domination and thus the inclusion of pluralist elements has enabled rather than prevented further solidarisation.

The core element that brings solidarisation and impartiality back in is the long-term emission reduction goal and the review/ambition mechanism, which builds the cornerstone of the PA. Article 4 establishes a mechanism which puts the Nationally Determined Contributions (NDCs) under international review. The PA obliges states to submit NDCs on a regular basis, i.e. every five years (Article 4.9). The submissions are put on public record (Article 4.12; UNFCCC 2017). Furthermore, the ambition mechanism requires that successive NDCs 'will represent a progression [...] and reflect [a Party's] highest possible ambition' (Article 4.3). This mechanism is extremely significant because it ensures that 'backsliding from past pledges is explicitly outlawed' (Bergamaschi et al., 11). Moreover, Article 15 establishes a mechanism to facilitate the implementation of a review system. This mechanism largely consists of an expert-based committee. With Article 15, the PA clearly addresses the issue of compliance, but does so in a cautious way in order not to endanger overall support (Voigt 2016, 162).

The review/ambition mechanism constitutes a solidarising effect according to several indicators: first, it exposes national sovereignty to the scrutiny of the international community. States are formally entitled to take a fully sovereign decision about their respective contribution to global GHG reduction. There are no specifications about how detailed particular types of action have to be stated, allowing relatively great latitude to each state. On the other hand, 'it was virtually impossible to refuse to submit [NDCs]' (Keohane and Oppenheimer 2016, 146). And the unanticipated number of submissions prior to the Paris summit (Keohane and Oppenheimer 2016, 146; Kinley 2016, 2) testifies to the fact that states indeed felt bound by the core idea of the new regime setup, even though their commitment in sum will not suffice to stick to the two-degree goal. NDCs have to be submitted by Parties on a regular basis and with an increasing degree of ambition, which further restricts the sovereign decision-making authority of states.

To put it simply, states do not have the right to reduce their commitment over time, and they are even obliged to raise and upgrade their activities gradually. That 'peer-pressure' and 'naming-and-shaming' are the only sanctions for non-compliance, may soften the solidarist and impartial character, but does not repeal it. The public record of NDCs constitutes a solidarising element because it clearly exposes states' decisions to the international community's and the public's scrutiny. The solidarising effect would have been even greater if the PA had also determined that the national pledges shall be *assessed* systematically rather than just published, and if it had codified sanctions for non-compliance. Such an individual assessment would have provided for greater transparency, but was very much contested, because 'major developing countries [...] perceived this system as infringing their national sovereignty' (Brun 2016, 119, see also *Chapter 5*). Behind this perception clearly lies a claim to justice as non-domination and the weak variant of mutual recognition. Consequently, instead of individual assessments, Article 14 of the PA stipulates a system of regular global stocktaking in order to assess collective progress towards effective implementation and towards achieving the long-term two- or 1.5-degree goal. The global stocktaking is supposed to take place 'two years ahead of each new national pledging cycle' (Brun 2016, 119). A direct review of national pledges would have been an even stronger incident of solidarisation, but the combination of collective tracking with public record of national pledges, is still considerable.

In sum, turning away from its initial preference of a strong, top-down, regulated climate change regime, with the PA the EU has certainly diminished some dimensions of its solidarist agenda and at the same time loosened up its focus on measures strengthening impartiality. As argued before, a top-down regime with strong regulations and effective sanctioning mechanisms would have enhanced solidarism in terms of the depth dimension of cooperation. And, it would have entailed stronger claims of justice as impartiality. However, promoting preferences for strong, regulating approaches more cautiously has ensured even greater levels of solidarisation in terms of the scope dimension of cooperation. The Copenhagen summit had made it crystal clear that the EU and other promoters of a strong climate regime, which mainly focusses on principles in line with impartiality, might quickly end up achieving nothing. Taking into account pluralist reservations – perhaps counter-intuitively – secured further solidarisation in terms of keeping more actors involved in the regime and at the same time left open the option for further solidarisation – including the depth of cooperation.

Similarly, shifting from a clear preference for impartiality towards including claims for non-domination and mutual recognition, at first sight curtails the significance of impartiality. At closer inspection, however, we find that allowing for greater relevance of mutual recognition and non-domination has been a precondition to adhere to a certain level of impartial justice claims in the regime. This is a fundamental ambiguity that the EU has faced as an international climate actor. These findings thus confirm our arguments in *Chapter 2* about the crucial necessity of finding a balance between all three conceptions of justice, i.e. the danger of moving too far into either corner of the justice triangle. While this does not mean that the right balance lies always in the middle, it underlines the importance of all three conceptions of justice and different political strategies that accommodate for solidarist and pluralist realities of the current order when it comes to building a comprehensive and just climate regime.

6.4 The implications of the solidarisation/climate justice nexus

In the previous section, we have combined a solidarisation perspective with a focus on the promotion of the different forms of procedural justice that have informed our analysis in this book, conceptualised as impartiality, non-domination, and mutual recognition. Bringing together these two approaches suggests an alternative reading of the EU's role in the climate change regime. Our analysis demonstrates that the strategic change of EU global climate policies in the aftermath of the Copenhagen summit is not merely a renunciation from its ambitious agenda of promoting a more solidarist version of international society as well as stronger claims for justice as impartiality. The relation between solidarism and pluralism in climate change on the one hand and the three different conceptions of climate justice on the other are much more complex and multi-layered to draw the simple conclusion that more solidarism and more impartiality per se would render the climate regime more advanced, more effective, more just.

The previous section has for instance revealed that solidarisation and justice as impartiality indeed bear striking similarities, such as their preference for universal solutions and principles and for the recognition of the rights and duties of individuals and other non-state actors. Both these aspects tend to challenge the primacy of states as international actors. We have argued that some degree of this solidarist impartiality is key to a climate change regime that claims serious prospects for success in the combat against global warming. Furthermore, the discussion of the CBDR has demonstrated that solidarisation and justice considerations can cause a trade-off. While from a solidarisation point of view, the CBDR is rather problematic, from a climate justice perspective there are good reasons to defend the CBDR as a core principle of the climate regime, even though the CBDR mainly speaks to claims of justice as non-domination and mutual recognition.

Most importantly, the discussion of the PA has brought to the fore that the 2016 climate treaty does indeed entail a partial reversion to pluralism and a stronger relevance of mutual recognition and non-domination at the expense of impartiality. However, rather than proving that impartiality and solidarist ambitions had been abandoned, our analysis suggests a new balance between solidarism and pluralism, and thus of the three justice conceptions. This changed balance constitutes much more than a mere abandonment of previous ambition. Instead, it was the means to re-introduce this very ambition through the back door. This is an expression and a consequence of the underlying ambiguity of pluralist and solidarist strands embedded in international society. In designing and promoting an advanced, effective, legitimate, and just climate regime, it is not possible to take an ultimate decision between solidarism and pluralism or for one of the justice conceptions at the complete expense of both the others.

Thus, combining both approaches and taking the underlying ambiguities into account bears two important consequences. First, it provides an alternative understanding of the EU's learning process in the climate change negotiations. The EU has not simply lowered its ambition in light of severe resistance towards its own preferred ideas of international order – the continued high share of arguments promoting impartiality up until 2019 in our content analysis in *Chapter 3* underlines this finding. In changing its strategy after 2009, the EU has taken into account fundamental underlying ambiguities of international order. Rather than just reducing ambition or giving up entirely on its preferred architecture of the regime, we interpret the EU's role as an attempt to accept and embrace the fundamental undecidability between the different legitimate claims towards a just international society in the context of climate change.

Second, this alternative reading of the EU and its role in the climate regime has certain implications for the concept of a (green) normative power and for normativity in international politics altogether. The most important implication is that promoting particular normative positions in international politics can never be exempt from normative downsides. There are very persuasive reasons (for the EU) to promote the solidarisation of the climate regime and the enhancement of justice as impartiality within it. However, this legitimate objective will inevitably do harm to other legitimate normative concerns. Given this fundamental ambiguity, it would be the wrong reaction to completely step back from any normative concerns and ambitions. In contrast, taking normative concerns into account requires an awareness of such ambiguities and of the potential flaws and drawbacks of one's own legitimate normative position. David Campbell's formulation provides an excellent summary of the requirements of a normative political actor:

> This rendering of ethics does not demand that a state's [we shall add: or the EU's] foreign policy be unambiguous and thoroughly consistent. All it requires is that an appreciation for ambiguity and a sensitivity to contingency be built into foreign policy.
>
> (Campbell 1993, 94)

This also puts our own claims as the authors of this book into perspective: rather than suggesting a particular normative standpoint as the ultimately preferable one, our claims are more *critical* than normative in character. In forming its political preferences and strategies, the EU – as well as any observer or analyst of international politics in assessing the normative qualities of international order – should acknowledge a certain inevitable degree of ambiguity and the need to balance different justice claims, including those brought forward by developing countries. Uncovering and embracing such ambiguities does not mean abandoning normative positions and ambitions; instead, it means to be self-reflective about one's own positioning.

Note

1 The AWG-KP was already in place since 2005, whereas AWG-LCA was newly established by the Bali Action Plan (UNFCCC (n.d.).

7 Conclusion

The EU, climate change, and balancing global justice

The starting point of this book was the widespread consensus in political and academic debates that the European Union (EU) has pursued progressive climate policies and was instrumental in the development of the global climate regime. This assessment is based on the EU's active involvement in global climate policy since the early 1990s and – at least in comparison to other large emitters, such as the United States (US), China, or Russia – its more ambitious internal measures, especially concerning binding emission reduction targets. This unique role of the EU in the climate regime has sparked an extensive research agenda on the effectiveness of the EU's climate policies and its role as a leader in the international climate negotiations (Oberthür and Roche Kelly 2008; Wurzel and Connelly 2011b; Bäckstrand and Elgström 2013; Wurzel, Connelly, and Liefferink 2017). The positive perception of EU global climate policy is closely tied to debates about its character as a normative power (Manners 2006; Whitman 2011). This has led some to ponder whether we can classify the EU as a 'green normative power' (Falkner 2007) that might be particularly able to promote environmental and climate protection in international society (Falkner and Buzan 2019).

Such a positive contribution would have to include questions of global justice within the climate regime. Yet only very few works have explicitly considered how justice relates to the EU's global climate policies. Given the broad literature on climate justice in general, and the countless justice questions related to climate change – among them, the uneven contribution to the problem and the variations in affectedness to name but two – (Gardiner et al. 2010; Heyward and Roser 2016), this gap is surprising. Thus, in this book we have analysed the ways in which the EU's global climate policy invokes justice concerns, implicitly or explicitly, how these concerns have shifted over time, to what effect, how we may explain these shifts, and how other key actors in the climate negotiations, especially large developing countries, perceive the EU and its changing climate policies. Our focus was on procedural justice (Eriksen 2016; Sjursen 2017), following the argument that substantive justice cannot be achieved without at least a considerable degree of procedural justice. In order to contextualise the developments of

EU climate justice policies and provide greater depth to our concepts, we have linked our narrative of the EU's involvement in the global climate regime to its efforts to solidarise international society and discussed the linkage between justice claims and securitisation.

7.1 The relevance of procedural justice

As discussed in the *Introduction* and *Chapter 2*, we have focussed on procedural instead of substantive justice, although the latter dominates the existing climate justice literature (e.g. Meyer and Roser 2010; Page 2013; Caney 2014). We think that our approach better captures the non-ideal political circumstances in the climate field and is also better suited to analyse political negotiations and their outcomes. The insights from the book's empirical chapters have corroborated this argument. Through their use of non-domination, impartiality, and mutual recognition as three conceptions of and approaches to procedural justice, they have identified key transformations and central points of contestation between different actors that exceed the debates about fair distribution of emission allowances or costs, and instead particularly pertain to the political rules of the game (Eriksen 2016).

Looking at the EU's involvement in the climate negotiations from the perspective of the EU and the BASIC countries (Brazil, South Africa, India, and China), respectively, *Chapters 3* and *5* have shown that key points of contestation have related not only to specific emission reduction targets but also to the organisation of the negotiations themselves and the political and economic architecture of the prevailing international order. Especially in the 1990s and 2000s, the EU was actively participating in building a liberal-cosmopolitan international society with an emphasis on global governance structures that it considered superior to the old pluralist alternatives, and the centrality they had afforded to the sovereignty of individual states. The new liberal order of a solidarist international society emphasised individual human rights, international law, and supranational institutions, and challenged the role of states as the only constituents of international society. Within that context, the EU climate policies and its preferences for the structure of the climate regime, which heavily focussed on measures we have linked to justice as impartiality, clearly mirrored the belief that such an order would deliver more effective and ultimately just solutions to cope with climate change. Yet its focus on individualism and a global liberal market order at the same time privileged particular conceptions of justice and cosmopolitanism over others (Ngcoya 2015).

Moreover, the EU from the very beginning displayed a strong belief in climate science and the ability to overcome the problem of climate change by implementing the right technical and regulatory decisions. Thus, the EU displayed a strong faith in international institutions such as the Intergovernmental Panel on Climate Change (IPCC) and the United Nations Framework Convention on Climate Change (UNFCCC), and perceived them as impartial and able to deliver climate regulations that would benefit all countries and people, and effectively mitigate climate change. As *Chapter 4* shows, the EU legitimised its proposals by a series of securitising moves, which framed climate change as an urgent global threat demanding universal action and binding rules. Eventually, these efforts paid off when the Kyoto Protocol was adopted, which, despite its limits and flexible mechanisms, in many ways mirrored the EU's understanding of how the climate regime should be organised and what it should aim at.

In contrast, many developing countries and especially the increasingly more influential BASIC countries had a very different impression of the climate negotiations and its outcomes. For them, the prevailing international order and the embedded climate institutions – including climate science, the IPCC, and the UNFCCC – were far from impartial. While not rejecting science or the need to build global institutions as such, they saw the concrete structures of the regime and the processes through which they were agreed at least partly as intertwined with colonial legacies and dominated by powerful developed countries that were now trying to pass on their climate debt to the developing world. Developing countries thus perceived their sovereignty to be threatened by top-down and binding climate treaties and regulations because they consider the international order to be structurally unequal and working to their disadvantage. For them, climate change and climate justice are primarily political problems that cannot be solved by scientific expertise, supranational institutions, and technocratic emission reduction targets alone. Thus, the EU's persistence on measures strengthening impartiality and especially its focus from 2007 onwards on gradually having large developing countries committing to binding emission reduction targets has been seen as an attempt to deflect from the historical responsibility of developed countries and a continuation of past imperial projects, possibly leading to a new 'green' colonialism (Huggan and Tiffin 2007; Lyons and Westoby 2014).

Consequently, large developing countries favoured bottom-up and voluntary approaches that would prevent breaches with national sovereignty and give them time to economically catch up with the Global North, i.e. realise their 'right to develop' (Moellendorf 2015, 178). They also have strongly defended their interpretation of the principle of Common But Differentiated Responsibilities (CBDR), which stands in contrast to top-down, binding targets for all countries, whereas, as we discuss in *Chapter 3*, the EU has always understood CBDR within the broader framework of impartiality. Instead, as *Chapter 5* has shown, developing countries, and in particular the BASIC countries, have mainly favoured principles that would strengthen justice as non-domination and mutual recognition.

Many developing countries were also discontent with how the EU – primarily through its member state Denmark – conducted the negotiations at the fifteenth Conference of the Parties of the UNFCCC (COP-15) in Copenhagen in 2009. Eventually, these very different perspectives on the *procedural* aspects of the climate regime – and not so much a disagreement over *substantive* climate justice principles i.e. what should be done to stop climate change, who should reduce by what percentage or who should pay what amount – played a key role in the failure to reach an agreement in Copenhagen. As elaborated in the next section, understanding these procedural aspects was also instrumental for making sense of the changes in the EU's climate policies, its negotiation strategies, and the organisational changes in the climate regime between the COPs in Copenhagen in 2009 and Paris in 2015.

Haying said that, our approach is not purely procedural (Brandstedt and Brülde 2019). Especially the conception of justice as impartiality and its emphasis on universal standards is intertwined with (often Western) substantive assumptions (Augenstein 2019) and partly overlaps with the principles discussed in the existing literature on climate justice (Caney 2010). However, our analysis shows that balancing impartiality with other procedural justice conceptions and linking it to concrete political categories (supranational institutions and top-down organisation of obligations, legal bindingness), enhances its empirical applicability. Our approach thus enables bridge building between the ideal typical assumptions about climate justice and the large International Relations and EU Studies climate policy literature, while contributing empirically as well as theoretically to non-ideal climate justice theory.

7.2 Understanding the shift from impartiality to non-domination and mutual recognition

Empirically, we have identified a crucial shift in the policies of the EU, the international negotiations, and the climate agreements from COP-15 to COP-21. As all chapters underline, the attention has increasingly shifted from top-down, binding emission reduction targets, which we link to impartiality, towards voluntary, bottom-up measures and attempts to acknowledge and integrate different perspectives and capabilities, which we see as strengthening non-domination and mutual recognition.

Analysing the EU, we find that this transformation is partly caused by internal institutional changes like the EU enlargement, the strengthening of the European External Action Service (EEAS) and the Green Diplomacy Network (GDN), and changes in the influence of different EU institutions. Beyond this, however, changed external circumstances have also played an important role in pushing the EU to reconsider its approach. Key examples are the rising Greenhouse Gas (GHG) emissions and increased political influence of large developing countries, the 2007–08 financial crisis, the demise of the liberal order and a re-emergence of the role of nation states. Overall, we identify a process of strategic learning in parts of the EU climate policy circles, especially in Directorate-General for Climate Action (DG CLIMA) and the EEAS. After the failure to anchor its preferred regime architecture in COP-15, these actors realised that the EU had to strengthen its climate diplomacy and also accept different approaches to climate change abatement if it wanted to stay relevant to the negotiations and succeed in adopting a new, more comprehensive climate agreement in 2015.

Having said that, our qualitative content analysis of key EU climate documents in *Chapter 3* (see *Figure 3.1*) also reveals that despite a gradual decrease in importance, the EU still favours instruments which we link to justice as impartiality. Thus, although it has strategically adapted its external climate policy and negotiation behaviour, its ultimate goal remains a top-down, binding regime architecture with mitigation targets for large emitter states (Interview 2018c). Consequently, the EU has striven to re-reintroduce impartiality 'through the backdoor' in the negotiation rounds following Paris (see *Chapter 6* and especially *Section 7.4*). This included the strengthening of the transparency and reporting rules, and at the Madrid COP in 2019 went as far as refusing the adoption of a deal instead of accepting too weak rules.

Beyond impartiality, the EU seems to genuinely welcome measures we would link to mutual recognition, e.g. recognising different perspectives like those from the Global South (including concepts such as 'ownership' and 'local knowledge'), integrating civil society, increasing transparency, and strengthening dialogical structures (e.g. the Talanoa Dialogue Platform agreed at COP-21). We have found this preference not only in our content analysis, with a rise towards the end of the research period, but also in EU climate policies and here especially in its climate development policies (von Lucke 2020a; GCCA+ 2018b). However, the EU remains reluctant to move towards measures that emphasise non-domination, especially if this entails a re-emergence of statist structures, national sovereignty and merely voluntary emission reduction targets, but less so in relation to market mechanisms, which have been playing an important role in the EU's own climate policies. These preferences concerning non-domination are heavily intertwined with the nature and role of the EU itself, which tries to transcend traditional Westphalian structures (Ahrens 2019), but at the same time has its roots in economic cooperation and trade. Thus, allowing states to play a more central role in the negotiations and to weaken international institutions and international law would go against the very idea of the EU itself and also against the world order it would like to create.

And yet, as *Chapter 6* has demonstrated, it formed part of the EU's learning process to accept that catering to claims of national sovereignty at least to some degree is necessary in order to further solidarist conceptions of order. Due to fundamentally changing circumstances, such as a general weakening of the liberal moment in world politics, a decisively solidarist agenda in climate politics that disregards pluralist reservations is mostly perceived as self-righteous rather than as a model to follow suit. A firm, inflexible stance on top-down regulations and strong institutional enforcement measures, as pursued by the EU up until the Copenhagen summit, thus reflects a very principled approach, but was doomed to failure. Being more flexible on its own preferred principles enabled the EU to re-introduce these principles through the backdoor and thus to keep them on the table. Due to its own ambiguous structure as an entity that consists of sovereign nation states while at the same time trying to overcome national sovereignty, the EU is predestined to balance these poles and thus to embrace ambiguity (Ahrens 2018a). This ambiguity perspective on EU climate policies puts the shift from impartiality to non-domination and mutual recognition into perspective.

Whereas the change in EU strategies towards mutual recognition and non-domination has mainly been a strategic decision without abandoning the underlying impartial aims (see *Figure 3.1*, as well as *Chapter 6* and *Section 7.4*), the transformation of the climate regime between the Kyoto Protocol and the Paris Agreement (PA) is more fundamental. The PA keeps some aspects that still strengthen impartiality, e.g. the aspiration to limit global warming to two or even 1.5 degrees Celsius and the longterm emission reduction targets. However, its central organising principle, the Nationally Determined Contributions (NDCs), clearly strengthen non-domination in their voluntariness and the concept of leaving it to individual member states to decide their own targets, as well as mutual recognition in the acknowledgement of different capabilities, the bottom-up structure, and transparency rules. We find that the EU's change in strategy contributed to this transformation. Its diplomatic outreach activities during the years preceding Paris, its efforts in forming the High Ambition Coalition (HAC) and the effective preparation and organisation of COP-21 by EU member state France were important for avoiding some of the mistakes of COP-15 and regaining the trust of many developing countries. In addition, as *Chapter 4* exemplifies, the EU considerably reduced its securitisation moves after the failure of Copenhagen, and thus avoided to offensively promote principles in line with impartiality. Yet, besides these changes in the EU's approach, other factors, and especially the increasing influence of key actors with understandings of global justice different from the EU's have also played important roles.

Already in the years preceding the 2009 summit in Copenhagen, it became clear that the US and large developing countries were not willing to accept a mainly impartial regime architecture in global climate governance. Consequently, if one wanted to keep the negotiations on track and eventually come to a more comprehensive global agreement that would go beyond a small 'club' of countries such as the second phase of Kyoto, a deviation from the goal of impartiality and a transformation of the regime architecture was necessary. Particularly for the BASIC countries, a shift away from top-down, binding emission reduction targets was a core prerequisite for participation. It was also important for the BASIC countries that the EU and the international community would trust them to take appropriate mitigation action in line with their capabilities and development needs, yet without coercive measures. The non-binding character of the NDCs and the PA in general was also instrumental in securing the signature of the US because Congress would not have accepted a more binding international treaty. Finally, the changes between Kyoto and the PA also mirror broader transformations of international society. As *Chapter 6* has explored in detail, the liberal moment and the drive towards more solidarisation in the international society of the post-Cold War era, which included an attempt to increase impartiality not only in the climate regime, has come to an end. In the 2000s and especially 2010s, states have regained influence and have strengthened the pluralist aspects of international society. There has been a growth of right-wing populist parties and governments and a re-emergence of concerns for state sovereignty in many countries in Europe as well as globally, including the Global South. The elections of Donald Trump, Jair Bolsonaro, Victor Orbán, Narendra Modi, and the Brexit vote exemplify this transformation and have made it much harder to reach agreement on top-down, supranational, and binding climate change abatement provisions.

This book's empirical analyses have shown that a non-essentialist, dynamic conceptualisation of global justice can be a useful analytical tool for tracing and understanding key changes in the climate regime. Beyond that, such a conceptualisation can also help us to come to a more nuanced normative discussion, as the following section explains in more detail. Yet our approach also has its limits. The analyses have shown that the global reach and complexity of the climate regime make it difficult to clearly pinpoint causal mechanisms. Thus, while we think that both the EU's changed approach to the negotiations as well as transformations in the positions and influence of other key actors were instrumental for the changing regime architecture between Kyoto and Paris, we cannot with absolute certainty say which factors played the more important role or might have incited others but rather see them as mutually reinforcing. Moreover, by focussing on global justice and on states and the EU, we have largely excluded other important explanatory variables such as economic considerations, the role of business actors as well as the influence of social movements or new scientific discoveries about climate change. Nevertheless, we think that our approach helps to better understand key aspects of the climate negotiations and also contributes to a more detailed and critical picture of central actors.

7.3 The normative dimension of global justice: finding a balance

Besides its analytical value, our tripartite framework of global justice also helps us to discuss the normative implications of the transformations in the climate regime and EU climate policy. As already briefly discussed in *Chapter 2*, the normative yardstick for this book is the 1.5-degree target and the prevention of human suffering along with the appropriate involvement of vulnerable actors in the climate regime. We argue that we need a balance between all three approaches to global justice in order to realise political solutions to these ends. So how do we assess the changes between Kyoto and Paris in relation to such a balance?

From the perspective of (green) normative power Europe, which often coincides with the perspective of the EU, the decrease in impartiality and the move towards non-domination and mutual recognition constitute a lowering of ambition, some might even say a capitulation in the light of an increasingly pluralist international society. However, to keep the climate regime within the UN framework and thus ensure its global reach and in order to adhere to some solidarist endeavours, these changes were unavoidable. In addition, a realignment of the EU's narrow focus on its own, regulatory, top-down approach was necessary for maintaining the EU's role as an important climate actor both EU-internally and in diplomatic settings. Given Europe's long colonial history, often forgotten within the EU but certainly not in the once colonised states and communities, the sometimes self-righteous and often tone-deaf and missionary behaviour of the EU in the climate negotiations was ill-received by many, especially large developing countries. The EU strategy to split the Group of 77 (G77) and its attempts to bring large developing countries to adopt binding mitigation targets were seen as evading its own responsibilities and as an encroachment on sovereignty.

As highlighted in *Chapter 4*, a too strong and narrow focus on impartiality combined with intense securitisation moves has the potential of aiding a (neo-)imperialist agenda and can severely narrow down the discursive space by ridiculing opposing arguments and different perspectives. It has also been used as an excuse to push the expansion of specific industrial sectors, e.g. the 'green economy', to the benefit of EU companies (Roeck, Delputte, and Orbie 2016; Roeck 2019; von Lucke 2020a) or to deflect from the historical responsibility of the Global North and the necessary economic transformations in developed countries. In combination with the in fact never truly impartial international institutions in the climate regime, the continuous power inequalities in international society, and the often exclusionary tendencies during the climate negotiations vis-à-vis actors from the Global South, a rethinking of the EU's approach and of impartiality in general was necessary. Thus, from a normative perspective, we think that especially the move towards mutual recognition was long overdue. It can help to strengthen what others have termed a 'pluriverse' (Escobar 2018) or 'polycentric' framework (Ostrom 2010). Recognising the legitimate concerns of countries and people from the Global South, who are least responsible for climate change but at the same time most vulnerable to its negative effects, is vital to reach a fair and ultimately also effective agreement. Developed countries, in general, and the EU, in particular, had to acknowledge that climate change is not merely a technical or economic problem but instead deeply political and interwoven with past and present inequalities that have to be addressed simultaneously. Moreover, reforming the EU's negotiation strategy before and during the COPs was also essential to growing support for its policy aims (Walker 2018) and to increasing the likelihood that the agreed measures will actually be implemented domestically.

This being said, we think that the move towards mutual recognition is incomplete. As we have already briefly discussed in *Chapter 2* and in the empirical analysis, despite an increase in Non-Governmental Organisations (NGOs) participation in COPs (Nasiritousi, Hjerpe, and Linnér 2016; Thew, Middlemiss, and Paavola 2020), their direct influence remains limited and states remain the principal actors in the climate regime, which at best amounts to a weak form of mutual recognition. In order to strengthen (strong) mutual recognition, the emphasis would have to be redirected to a variety of non-state actors. The role of the Fridays for Future (FFF) movement in raising attention for climate justice concerns, and also its influence on the 2019 European elections in some member states and in pushing the EU towards the Green Deal (DW 2019), exemplify that a better recognition of civil society demands can improve climate policy and benefit climate justice. However, such recognition needs to go beyond well-known NGOs and Northern activists and particularly include indigenous peoples, social movements, local grassroots organisations, and activists especially from the Global South and from those regions particularly vulnerable to climate change. This would not only enhance the recognition of the ones most affected by climate change but could also strengthen the attention for the specific needs of minorities and other marginalised groups, whose already existing struggles will be exacerbated not only by the effects of climate change itself but sometimes also by misguided climate change abatement measures (Dehm 2016; Astuti and McGregor 2017; Fjellheim and Carl 2020).

Practically, this would mean to provide institutionalised procedures for these actors to voice their concerns and needs at the EU level – for instance in relation to the EU's climate development policy (von Lucke 2020a) – before policies and political strategies are developed, and should also include feedback on which recommendations the EU takes up and which ones it does not (Interview 2018g). In this context, the EU should also work towards anchoring the international climate regime to rules that give other states incentives to increase transparency and the inclusion of the most vulnerable stakeholders in the domestic formulation of their NDCs. Likewise, the EU has to push for the introduction of best practices regarding transparency and civil society inclusion in organising and implementing future COPs and other key climate summits. Finally, while we think that strengthening mutual recognition in the above sense can contribute to keep global warming below 1.5 degrees and prevent human suffering, theoretically it is difficult to decide which actors and preferences should be included in the end. Whereas our suggestion to focus on civil society and on those that are particularly affected and vulnerable might seem straightforward, on the practical level it is difficult to select specific actors and decide which positions to follow in the end without referring to a background theory or assumption of justice, hence without what we would term justice as impartiality.

While the securitisation of climate change is necessary to keep the momentum and further develop the climate regime towards the two-degree goal, *Chapter 4* has demonstrated how different securitising moves invoke different justice claims. In order to strengthen mutual recognition, the EU ought to take care, on the one hand, not to present itself as the main object of climate threats and thus reinforce non-domination. On the other hand, rather than simply constructing the threat as global or towards future generations, it needs to recognise the diverse ways in which actors in different locales and social and economic contexts are affected by climate change. While this weakens the simplistic grammar of securitisation, at the same time it legitimises the differentiated involvement of diverse actors yet does compromise on the urgency of climate action.

When it comes to the strengthening of non-domination, our assessment is more cautions. We agree that it is important to strengthen less powerful states in the negotiations. Thus, supporting their often smaller delegations, building progressive alliances or bilateral co-operation can be imperative to halt climate change and eventually also support vulnerable populations. This also includes respecting the sovereignty of developing countries and accepting different, more voluntary or adaptation-focussed approaches, especially from countries in the Global South. As *Chapter 5* shows with a view to the BASIC countries, reducing domination can help to build trust in large developing countries and eventually lead to more ambitious climate policies than relying on binding top-down targets, which are often met with resistance. Moreover, with a view on the international order in general and the climate regime in particular, which are still dominated by countries of the Global North, exposing and challenging domination and working towards a more equal distribution of influence certainly is in order.

However, given the global reach of climate change and its negative effects, the dispersion of responsibility and problems of freeriding, too much non-domination in the global climate regime is problematic. Leaving it to states to decide what to do about the problem without strong rules and international institutions is highly susceptible to domestic political changes, as the announcement of Donald Trump to withdraw from Paris or the election of Jair Bolsonaro and the subsequent dismantling of Brazil's climate policy exemplify. It also runs the risk that other seemingly more immediate crises such as economic downturns quickly overshadow effective climate policies. The voices calling for a suspension of climate mitigation provisions in the economic reconstruction amid the 2020 COVID-19 crisis exemplify this danger (Die Welt 2020; Friedman 2020). Hoping that 'peer-pressure' and 'naming and shaming' (Falkner 2016) without any form of legal accountability will bring states to publish ever more ambitious NDCs and also follow up on them seems unlikely given the experience of the past. At the same time, doing too little to abate climate change, which is a likely consequence of too much emphasis on voluntary instruments and sovereign decisions by states, constantly increases the indirect domination of the most vulnerable people around the world. This creates new cleavages in the climate regime, for instance when large (also in terms of GHG emissions) and increasingly powerful developing countries such as China or India refuse to be bound by mitigation rules, while smaller developing countries and especially their poorest and most vulnerable people continue to suffer and to be dominated by rising GHGs. The most drastic example are the small island states, which will almost certainly become uninhabitable due to climate change.

In the end, we think that, without a considerable degree of impartiality, hence a strong supranational component and a clear strengthening of individual rights, a complex collective action problem such as climate change is very difficult to solve. Without some impartial aspirations, strengthening mutual recognition and non-domination will be insufficient to keep global warming as low as possible and prevent widespread human suffering. Nevertheless, recalling our triangle of global justice from *Chapter 2* (*Figure 2.1*), our findings also underline that we have to avoid extremes when it comes to conceptions of justice and take contestation and ambiguity seriously, both as 'social activity' and as a 'mode of critique' (Wiener 2017). Thus, the impartial approach must be constantly checked to avoid unilateral interpretations or imperial overreach, and needs to be combined with fundamental reforms of the political procedures that lead to the measures and the institutions that oversee their implementation. Only if these become much more inclusive and reflective, especially towards the Global South and civil society representatives from vulnerable groups, will they be accepted as approximating their impartial aspirations and be able to sustainably strengthen global climate justice.

Annex

List of analysed documents for Chapter 3

European Commission	
Year	*Document*
1988	'The Greenhouse Effect and the Community.' Commission work programme concerning the evaluation of policy options to deal with the 'greenhouse effect'. Communication to the Council, COM(88) 656 final, Brussels.
1991	A Community Strategy to limit Carbon Dioxide emissions and to improve energy efficiency. Communication from the Commission to the Council, SEC(91) 1744 final, Brussels.
1996	Communication from the Commission under the UN Framework Convention on Climate Change. Article 4.2. (b) (c) (d) and Article 12. COM(96) 217 final, Brussels.
1997	Communication from the Commission on environment and employment (Building a sustainable Europe). COM(97) 592 final, Brussels.
1997	Communication from the Commission: An overall view of energy policy and actions. COM(97) 167 final, Brussels.
1999	'Preparing for Implementation of the Kyoto Protocol.' Commission communication to the Council and the Parliament, COM(1999) 230, Brussels
2000	Communication from the Commission to the Council and the European Parliament on EU policies and measures to reduce greenhouse gas emissions: Towards a European Climate Change Programme (ECCP). COM(2000) 88 final, Brussels.
2003	Proposal for a decision of the European Parliament and of the Council for a monitoring mechanism of Community greenhouse gas emissions and the implementation of the Kyoto Protocol (presented by the Commission). COM(2003) 51 final, Brussels.
2005	Communication from the Commission to the Council, the European Parliament, the European Economic and Social Committee and the Committee of the Regions. Winning the Battle Against Global Climate Change. COM(2005) 35 final, Brussels.
2006	Green Paper. A European Strategy for Sustainable, Competitive and Secure Energy. COM(2006) 105 final, Brussels.
2008	Communication from the Commission to the European Parliament, the Council, the European Economic and Social Committee and the Committee of the Regions. 20 20 by 2020. Europe's climate change opportunity. COM(2008) 30 final, Brussels.

(Continued)

European Commission

Year	*Document*
2008	Boosting growth and jobs by meeting our climate change commitments. IP/08/80, Brussels.
2009	Communication from the Commission to the European Parliament, the Council, the European Economic and Social Committee and the Committee of the Regions. Towards a comprehensive climate change agreement in Copenhagen. COM(2009) 39 final, Brussels.
2010	Communication from the Commission to the European Parliament, the Council, the European Economic and Social Committee and the Committee of the Regions. Analysis of options to move beyond 20 per cent greenhouse gas emission reductions and assessing the risk of carbon leakage. COM(2010) 265 final, Brussels.
2011	Communication from the Commission to the European Parliament, the Council, the European Economic and Social Committee and the Committee of the Regions. A Roadmap for moving to a competitive low carbon economy in 2050. COM(2011) 112 final, Brussels.
2013	Communication from the Commission to the European Parliament, the Council, the European Economic and Social Committee and the Committee of the Regions. An EU Strategy on adaptation to climate change. COM(2013) 216 final, Brussels.
2013	Communication from the Commission to the European Parliament, the Council, the European Economic and Social Committee and the Committee of the Regions. The 2015 International Climate Change Agreement: Shaping international climate policy beyond 2020. Consultative Communication {SWD(2013) 97 final}. COM(2013) 167 final, Brussels.
2014	Communication from the Commission to the European Parliament, the Council, the European Economic and Social Committee and the Committee of the Regions. A policy framework for climate and energy in the period from 2020 to 2030. COM(2014) 15 final/2, Brussels.
2016	Proposal for a Council Decisions on the signing, on behalf of the European Union, of the Paris Agreement adopted under the United Nations Framework Convention on Climate Change. COM(2016) 62 final, Brussels.
2016	Communication from the Commission to the European Parliament and the Council. The Road from Paris: assessing the implications of the Paris Agreement and accompanying the proposal for a Council decision on the signing, on behalf of the European Union, of the Paris agreement adopted under the United Nations Framework Convention on Climate Change. COM(2016) 110 final, Brussels.
2018	Communication from the Commission to the European Parliament, the European Council, the Council, the European Economic and Social Committee, the Committee of the Regions and the European Investment Bank. A Clean Planet for all. A European strategic long-term vision for a prosperous, modern, competitive and climate neutral economy. COM(2018) 773 final, Brussels.

European Council

Year	*Document*
1990	Dublin European Council. Declaration by the European Council on the environmental imperative. Bull. EC 6–1990.
1997	Amsterdam European Council. 16 and 17 June 1997. Presidency Conclusions. Press release: 18.06.97.

European Council

Year	*Document*
2004	Brussels European Council. 25 and 26 March 2004. Presidency Conclusions. 9048/04, Brussels.
2005	European Council Brussels. 22 and 23 March 2005. Presidency Conclusions. 7619/1/05, Brussels.
2007	Brussels European Council. 8 and 9 March 2007. Presidency Conclusions. 7224/1/07, Brussels.
2007	Brussels European Council. 21 and 22 June 2007. Presidency Conclusions. 11177/1/07, Brussels.
2007	Conclusions. Energy policy for Europe. A contribution of the Energy Ministers to the 2007 Spring European Council. 5411/07, Brussels.
2009	Brussels European Council. 19 and 20 March 2009. Presidency Conclusions. 7880/09, Brussels.
2009	Brussels European Council. 29 and 30 October 2009. Presidency Conclusions. 15265/1/09, Brussels.
2010	Presidency Conclusions: climate finance. 16508/10, Brussels.
2011	Commission Communication on a Roadmap for moving to a competitive low carbon economy in 2050 – Presidency conclusions. 11964/11, Brussels.
2012	A Roadmap for moving to a competitive low-carbon economy in 2050-Presidency conclusions. 6842/12, Brussels.

Council of the European Union

Year	*Document*
1993	Council Decision of 15 December 1993 concerning the conclusion of the United Nations Framework Convention on Climate Change. L 33/11, Brussels.
1996	Press Release 96/188, Brussels.
1997	Council Resolution of 27 June 1997 on renewable sources of energy. No C 210/1, Brussels.
2002	Council Decision of 25 April 2002 concerning the approval, on behalf of the European Community, of the Kyoto Protocol to the United Nations Framework Convention on Climate Change and the joint fulfilment of commitments thereunder. L 130/1, Brussels.
2007	PRESS RELEASE 2785th Council Meeting. Environment. 6272/07 (Presse 25), Brussels.
2014	Conclusions (23 and 24 October 2014). EUCO 169/14, Brussels.
2015	PRESS RELEASE EU position for the UN climate change conference in Paris: Council conclusions. 657/15, Brussels.
2015	Council Decision (EU) 2015/1339 of 13 July 2015 on the conclusion, on behalf of the European Union, of the Doha Amendment to the Kyoto Protocol to the United Nations Framework Convention on Climate Change and the joint fulfilment of commitments thereunder. L 207/1, Brussels.
2018	Preparations for the UNFCCC meetings in Katowice (2–14 December 2018). Council conclusions. 12901/18, Brussels.
2019	Climate Diplomacy. Council conclusions. 6153/19, Brussels.

(Continued)

European Parliament	
Year	*Document*
1998	Legislative resolution embodying Parliament's opinion on the proposal for a Council Directive on transportable pressure equipment (COM(96)0674 – C4–0068/97 – 97/0011(SYN)). C 80/217, Brussels.
1998	Resolution on climate change in the run-up to Buenos Aires (November 1998). C 313, Brussels.
2001	European Union strategy for the Marrakesh Conference on Climate Change. B5–0686/2001.European Parliament resolution on the follow-up to Parliament's opinion on the European Union's strategy for the Marrakesh Conference on Climate Change. C 112 E/299, Brussels.
2002	Outcome of the Bonn Conference on climate change. B5–0539, 0540, 0541, 0543, 0551, and 0552/2001. European Parliament resolution on the outcome of the Bonn Conference on climate change. C 72 E/321, Brussels.
2003	Directive 2003/87/EC of the European Parliament and of the Council of 13 October 2003 establishing a scheme for greenhouse gas emission allowance trading within the Community and amending Council Directive 96/61/EC. L 275/32, Brussels.
2006	Nairobi Conference on Climate Change. European Parliament resolution on the European Union strategy for the Nairobi Conference on Climate Change (COP 12 and COP/MOP 2). C 313 E/439, Brussels.
2007	Climate change. European Parliament resolution on climate change. P6_TA(2007)0038, Brussels.
2009	2050: The future begins today – recommendations for the EU's future integrated policy on climate change European Parliament resolution of 4 February 2009 on '2050: The future begins today – Recommendations for the EU's future integrated policy on climate change' (2008/2105(INI)). P6_TA(2009)0042, Brussels.
2009	Copenhagen Conference on Climate Change European Parliament resolution of 25 November 2009 on the EU strategy for the Copenhagen Conference on Climate Change (COP 15). P7_TA(2009)0089, Brussels.
2009	Decision No 406/2009/EC of the European Parliament and of the Council of 23 April 2009 on the effort of Member States to reduce their greenhouse gas emissions to meet the Community's greenhouse gas emission reduction commitments up to 2020. L 140/136, Brussels.
2009	Directive 2009/28/EC of the European Parliament and of the Council of 23 April 2009 on the promotion of the use of energy from renewable sources and amending and subsequently repealing Directives 2001/77/EC and 2003/30/EC. L 140/16, Brussels.
2013	Regulation (EU) No 525/2013 of the European Parliament and of the Council of 21 May 2013 on a mechanism for monitoring and reporting greenhouse gas emissions and for reporting other information at national and Union level relevant to climate change and repealing Decision No 280/2004/EC. L 165/13, Brussels.
2018	Directive (EU) 2018/410 of the European Parliament and of the Council of 14 March 2018 amending Directive 2003/87/EC to enhance cost-effective emission reductions and low-carbon investments, and Decision (EU) 2015/1814. L 76/3, Brussels.

References

Aamodt, Solveig. 2018a. 'Environmental Ministries as Climate Policy Drivers: Comparing Brazil and India.' *The Journal of Environment & Development* 27 (4): 355–81.

Aamodt, Solveig. 2018b. 'The Ability to Influence: A Comparative Analysis of the Role of Advocacy Coalitions in Brazilian Climate Politics.' *Review of Policy Research* 35 (3): 372–97.

Aamodt, Solveig, and Elin L. Boasson. 2020. 'From Impartial Solutions to Mutual Recognition: Explaining Why the EU Changed Its Procedural Climate Justice Preferences.' *GLOBUS Research Paper* 3/2020.

Aamodt, Solveig, and Iselin Stensdal. 2017. 'Seizing Policy Windows: Policy Influence of Climate Advocacy Coalitions in Brazil, China, and India, 2000–2015.' *Global Environmental Change* 46: 114–25.

Abranches, Sergio. 2010. *Copenhague, Antes E Depois.* Rio de Janeiro: Civilização Brasileira.

Acharya, Amitav, and Barry Buzan. 2010. 'Why Is There No Non-Western International Relations Theory? An Introduction.' In *Non-Western International Relations Theory: Perspectives on and Beyond Asia*, edited by Amitav Acharya and Barry Buzan, 1–25. London: Routledge.

Adelle, Camilla, Katja Biedenkopf, and Diarmuid Torney, eds. 2018. *European Union External Environmental Policy: Rules, Regulation and Governance Beyond Borders.* The European Union in International Affairs. Cham: Palgrave Macmillan.

Agarwal, Anil, and Sunita Narain. 1991. *Global Warming in an Unequal World.* New Delhi: Centre for Science and Environment.

Agrawal, Arun, and Elinor Ostrom. 2001. 'Collective Action, Property Rights, and Decentralization in Resource Use in India and Nepal.' *Politics & Society* 29 (4): 485–514.

Aguilar-Støen, Mariel, and Cecilie Hirsch. 2015. 'Environmental Impact Assessments, Local Power and Self-Determination: The Case of Mining and Hydropower Development in Guatemala.' *The Extractive Industries and Society* 2 (3): 472–79.

Ahrens, Bettina. 2017. 'The Solidarisation of International Society: The EU in the Global Climate Change Regime.' *GLOBUS Research Paper* 5/2017.

Ahrens, Bettina. 2018a. 'AmbigEUity: The EU and the Solidarisation of International Society.' PhD Thesis, Tübingen. Accessed September 03, 2020. https://publikationen.uni-tuebingen.de/xmlui/handle/10900/83829.

Ahrens, Bettina. 2018b. 'Normative Power Europe in Crisis? Understanding the Productive Role of Ambiguity for the EU's Transformative Agenda.' *Asia Europe Journal* 16 (2): 199–212.

Ahrens, Bettina. 2019. 'The European Union between Solidarist Change and Pluralist Re-Enactment.' In *International Organization in the Anarchical Society: The Institutional Structure of World Order*, edited by Tonny Brems Knudsen and Cornelia Navari, 265–92. Palgrave Studies in International Relations. Cham: Springer International Publishing.

Ahrens, Bettina, and Thomas Diez. 2015. 'Solidarisation and Its Limits: The EU and the Transformation of International Society.' *Global Discourse* 5 (3): 341–55.

Alston, Margaret. 2014. 'Gender Mainstreaming and Climate Change.' *Women's Studies International Forum* 47: 287–94.

Ananthakrishnan, G. 2015. 'BASIC Optimistic About Legally Binding Paris Deal.' *The Hindu*, December 9. Accessed November 24, 2020. https://www.thehindu.com/sci-tech/energy-and-environment/cop21-basic-group-says-legally-binding-paris-deal-in-sight/article7962405.ece.

Anderson, Joel, and Axel Honneth. 2005. 'Autonomy, Vulnerability, Recognition, and Justice.' In *Autonomy and the Challenges to Liberalism: New Essays*, edited by John P. Christman and Joel Anderson, 127–49. Cambridge: Cambridge University Press.

Araya, Monica. 2015. 'FEATURE: The Role of Non-State Actors Will Grow After the Paris Climate Conference." Accessed March 23, 2017. https://cdkn.org/2015/07/growing-role-non-state-actors/?loclang=en_gb.

Arias Cañete, Miguel. 2016. 'How Will the Paris Agreement Impact EU Climate and Energy Policies?' Accessed August 20, 2020. http://www.investineu.com/blog/page/33/.

Armstrong, David. 2006. 'The Nature of Law in an Anarchical Society.' In *The Anarchical Society in a Globalized World*, edited by Richard Little and John Williams, 121–40. Basingstoke: Palgrave Macmillan.

Ashley, Richard K. 1988. 'Untying the Sovereign State: A Double Reading of the Anarchy Problematique.' *Millennium – Journal of International Studies* 17 (2): 227–62.

Ashley, Richard K. 1989. 'Imposing International Purpose: Notes on a Problematic Governance.' In *Global Changes and Theoretical Challenges: Approaches to World Politics for the 1990s*, edited by Ernst O. Czempiel and James N. Rosenau, 251–90. Lexington, MA: Lexington Books.

Ashley, Richard K., and R.B.J. Walker. 1990. 'Conclusion: Reading Dissidence/Writing the Discipline: Crisis and the Question of Sovereignty in International Studies.' *International Studies Quarterly* 34 (3): 367–416.

Astuti, Rini, and Andrew McGregor. 2017. 'Indigenous Land Claims or Green Grabs? Inclusions and Exclusions Within Forest Carbon Politics in Indonesia.' *The Journal of Peasant Studies* 44 (2): 445–66.

Augenstein, Lea. 2019. 'The Western Argument Wins: A Postcolonial Critique of Conceptions of Global Justice as Mutual Recognition.' GLOBUS Report 3.

Babbitt, Eileen F. 2011. 'Conflict Resolution and Human Rights: The State of the Art.' In *The Sage Handbook of Conflict Resolution*, edited by Jacob Bercovitch, Ira W. Zartman, and Viktor A. Kremenjuk, 613–29. Los Angeles, CA: Sage.

Bachvarova, Mira. 2013. 'Non-Domination's Role in the Theorizing of Global Justice.' *Journal of Global Ethics* 9 (2): 173–85.

Bäckstrand, Karin, and Ole Elgström. 2013. 'The EU's Role in Climate Change Negotiations: From Leader to "Leadiator".' *Journal of European Public Policy* 20 (10): 1369–86.

Bäckstrand, Karin, Jonathan W. Kuyper, Björn-Ola Linnér, and Eva Lövbrand. 2017. 'Non-State Actors in Global Climate Governance: From Copenhagen to Paris and Beyond.' *Environmental Politics* 26 (4): 561–79.

Bain, William. 2014. 'The Pluralist-Solidarist Debate in the English School.' In *Guide to the English School in International Studies*, edited by Cornelia Navari and Daniel Green, 159–69. Hoboken, NJ: Wiley-Blackwell.

Bals, Christoph. 2009. 'Klimazug im „Tal des Todes" zwischen Posen und Kopenhagen. Ergebnisse des UN-Klimagipfels in Polen, 1. -12. Dezember 2008.' Accessed March 03, 2018. http://www.germanwatch.org/klima/cl4res.pdf.

Balzacq, Thierry. 2011. 'A Theory of Securitization: Origins, Core Assumptions, and Variants.' In *Securitization Theory: How Security Problems Emerge and Dissolve*, edited by Thierry Balzacq, 1–30. PRIO New Security Studies. London: Routledge.

Barnes, Pamela M. 2011. 'The Role of the Commission of the European Union: Creating External Coherence from Internal Diversity.' In *The European Union as a Leader in International Climate Change Politics*, edited by Rüdiger Wurzel and James Connelly, 41–57. Routledge/UACES Contemporary European Studies 15. London: Routledge.

Barnett, Jon, and John Campbell. 2015. *Climate Change and Small Island States: Power, Knowledge and the South Pacific.* London: Routledge.

Barnett, Jon, Richard A. Matthew, and Karen O'Brien. 2010. 'Global Environmental Change and Human Security: An Introduction.' In *Global Environmental Change and Human Security*, edited by Richard A. Matthew, Jon Barnett, Bryan McDonald, and Karen O'Brien, 3–32. London: MIT Press.

Barrera, Julia, and Reimund Schwarze. 2004. 'Does the CDM Contribute to Sustainable Development? Evidence from the AIJ Pilot Phase.' *International Journal of Sustainable Development* 7 (4): 353–68.

BASIC. 2013. 'Joint Statement Issued at the Conclusion of the 15th Basic Ministerial Meeting on Climate Change.' Accessed August 20, 2020. https://www.environment.gov.za/jointstatement_15thbasicministerialmeeting.

Bellamy, Alex J. 2010. *Responsibility to Protect: The Global Effort to End Mass Atrocities.* Reprint. Cambridge: Polity.

Bellamy, Richard. 2013. '"An Ever Closer Union Among the Peoples of Europe": Republican Intergovernmentalism and Democratic Representation within the EU." *Journal of European Integration* 35 (5): 499–516.

Bergamaschi, Luca, Nick Mabey, Jonathan Gaventa, and Camilla Born. 'EU Foreign Policy in a Changing Climate: A Climate and Energy Strategy for Europe's Long-Term Security.' Accessed February 16, 2018. https://www.e3g.org/docs/E3G_EU_foreign_policy_energy_climate.pdf.

Biderman, Rachel (Observatório do Clima). 2009. 'Diretrizes para formulacão de políticas públicas em mudanças climáticas no Brasil.' São Paulo. Accessed November 24, 2020. https://ipam.org.br/wp-content/uploads/2009/05/diretrizes_para_formulac%CC%A7a%CC%83o_de_poli%CC%81ticas_.pdf.

Biedenkopf, Katja. 2016. 'Transatlantic Climate Diplomacy.' Accessed July 13, 2017. www.aicgs.org/publication/transatlantic-climate-diplomacy.

Blomfield, Megan. 2013. 'Global Common Resources and the Just Distribution of Emission Shares.' *Journal of Political Philosophy* 21 (3): 283–304.

Blomfield, Megan. 2016. 'Historical Use of the Climate Sink.' *Res Publica* 22 (1): 67–81.

Boasson, Elin L., and Jørgen Wettestad. 2016. *EU Climate Policy: Industry, Policy Interaction and External Environment.* London: Routledge.

Bodansky, Daniel. 2011. 'W[h]ither the Kyoto Protocol? Durban and Beyond', *Harvard Project on Climate Agreements.* Accessed November 30, 2020. https://papers.ssrn.com/sol3/papers.cfm?abstract_id=1917603.

Bode, Sven. 2004. 'Equal Emissions Per Capita Over Time – A Proposal to Combine Responsibility and Equity of Rights for Post-2012 GHG Emission Entitlement Allocation.' *European Environment* 14 (5): 300–16.

Booth, Ken. 1991. 'Security and Emancipation.' *Review of International Relations* 17 (4): 313–26.

Bourguignon, Didier. 2016. 'The Precautionary Principle: Definitions, Applications and Governance.' Accessed October 10, 2017. http://www.europarl.europa.eu/RegData/etudes/IDAN/2015/573876/EPRS_IDA(2015)573876_EN.pdf.

Brandstedt, Eric. 2019. 'Non-Ideal Climate Justice.' *Critical Review of International Social and Political Philosophy* 22 (2): 221–34.

Brandstedt, Eric, and Bengt Brülde. 2019. 'Towards a Theory of Pure Procedural Climate Justice.' *Journal of Applied Philosophy* 36 (5): 785–99.

Brugnach, Marcela, Marc Craps, and Art Dewulf. 2017. 'Including Indigenous Peoples in Climate Change Mitigation: Addressing Issues of Scale, Knowledge and Power.' *Climatic Change* 140 (1): 19–32.

Brun, Aslak. 2016. '"Conference Diplomacy": The Making of the Paris Agreement.' *Politics and Governance* 4 (3): 115–23.

Bulkeley, Harriet, and Heike Schroeder. 2012. 'Beyond State/Non-State Divides: Global Cities and the Governing of Climate Change.' *European Journal of International Relations* 18 (4): 743–66. http://ejt.sagepub.com/content/18/4/743.

Bull, Hedley. 1966. 'The Grotian Conception of International Society.' In *Diplomatic Investigations*, edited by Herbert Butterfield and Martin Wight, 51–73. London: Allen & Unwin.

Bull, Hedley. 1977. *The Anarchical Society: A Study of Order in World Politics.* Basingstoke: Macmillan.

Bull, Hedley. 2002. *The Anarchical Society: A Study of Order in World Politics.* 3rd ed. New York: Columbia University Press.

Burgess, J. P. 2011. *The Ethical Subject of Security: Geopolitical Reason and the Threat to Europe.* PRIO New Security Studies. London: Routledge.

Burns, Charlotte, and Neil Carter. 2011. 'The European Parliament and Climate Change: From Symbolism to Heroism and Back Again.' In *The European Union as a Leader in International Climate Change Politics*, edited by Rüdiger Wurzel and James Connelly, 58–73. Routledge/UACES Contemporary European Studies 15. London: Routledge.

Buzan, Barry. 2004. *From International to World Society? English School Theory and the Social Structure of Globalisation.* Cambridge: Cambridge University Press.

Buzan, Barry. 2006. 'Rethinking Hedley Bull on the Institutions of International Society.' In *The Anarchical Society in a Globalized World*, edited by Richard Little and John Williams, 75–96. Basingstoke: Palgrave Macmillan.

Buzan, Barry. 2014. *An Introduction to the English School of International Relations: The Societal Approach.* Cambridge: Polity Press.

Buzan, Barry, and Ole Wæver. 2004. *Regions and Powers: The Structure of International Security.* Cambridge: Cambridge University Press.

Buzan, Barry, Ole Wæver, and Jaap de Wilde. 1998. *Security: A New Framework for Analysis.* Boulder, CO: Lynne Rienner.

Campbell, David. 1993. *Politics Without Principle: Sovereignty, Ethics, and the Narratives of the Gulf War.* Critical Perspectives on World Politics. Boulder, CO: Lynne Rienner.

Caney, Simon. 2005. 'Cosmopolitan Justice, Responsibility, and Global Climate Change.' *Leiden Journal of International Law* 18 (4): 747–75.

Caney, Simon. 2006. 'Environmental Degradation, Reparations, and the Moral Significance of History.' *Journal of Social Philosophy* 37 (3): 464–82.

Caney, Simon. 2010. 'Climate Change and the Duties of the Advantaged.' *Critical Review of International Social and Political Philosophy* 13 (1): 203–28.

Caney, Simon. 2014. 'Two Kinds of Climate Justice: Avoiding Harm and Sharing Burdens.' *The Journal of Political Philosophy* 22 (2): 125–49.

Caney, Simon. 2016. 'The Struggle for Climate Justice in a Non-Ideal World.' *Midwest Studies in Philosophy* 40 (1): 9–26.

Capstick, Stuart B., and Nicholas F. Pidgeon. 2014. 'What is Climate Change Scepticism? Examination of the Concept Using a Mixed Methods Study of the UK Public.' *Global Environmental Change* 24: 389–401.

Caramel, Laurence. 2014. 'Besieged by the Rising Tides of Climate Change, Kiribati Buys Land in Fiji.' *The Guardian*, July 1. Accessed October 01, 2019. https://www.theguardian.com/environment/2014/jul/01/kiribati-climate-change-fiji-vanua-levu.

Carbonmarketwatch. 2017. 'Beyond the EU ETS: Strengthening Europe's Carbon Market through National Action.' *Carbon Market Watch Policy Briefing.* Accessed November 24, 2020. https://carbonmarketwatch.org/publications/beyond-eu-ets/.

Carlarne, Cinnamon P. 2010. *Climate Change Law and Policy: EU and US Perspectives.* Oxford: Oxford University Press.

Carlsson-Kanyama, Annika, and Alejandro D. Gonzalez. 2009. 'Potential Contributions of Food Consumption Patterns to Climate Change.' *The American Journal of Clinical Nutrition* 89 (5): 1704–9.

Chaban, Natalia, and Michèle Knodt. 2015. 'Energy Diplomacy in the Context of Multistakeholder Diplomacy: The EU and BICS.' *Cooperation and Conflict* 50 (4): 457–74.

Chandler, David. 2004. 'The Responsibility to Protect? Imposing the "Liberal Peace".' *International Peacekeeping* 11 (1): 59–81.

Chandler, David, Erika Cudworth, and Stephen Hobden. 2018. 'Anthropocene, Capitalocene and Liberal Cosmopolitan IR: A Response to Burke et al.'s "Planet Politics".' *Millennium – Journal of International Studies* 46 (2): 190–208.

Chhatre, Ashwini, and Arun Agrawal. 2009. 'Trade-Offs and Synergies between Carbon Storage and Livelihood Benefits from Forest Commons.' *PNAS* 106 (42): 17667–70.

Christoff, Peter. 2010. 'Cold Climate in Copenhagen: China and the United States at COP15.' *Environmental Politics* 19 (4): 637–56.

Christoff, Peter. 2016. 'The Promissory Note: COP 21 and the Paris Climate Agreement.' *Environmental Politics* 25 (5): 765–87.

Clark, Duncan. 2012. 'Has the Kyoto Protocol Made any Difference to Carbon Emissions?' *The Guardian*, November 26. Accessed November 04, 2016. https://www.theguardian.com/environment/blog/2012/nov/26/kyoto-protocol-carbon-emissions.

Clémençon, Raymond. 2016. 'The Two Sides of the Paris Climate Agreement.' *The Journal of Environment & Development* 25 (1): 3–24.

Climate Action Tracker. 2019. 'The CAT Thermometer: September 2019 Update.' Accessed October 15, 2019. https://climateactiontracker.org/global/cat-thermometer/.

Climate Home News. 2019. 'EU Aims for Long Term Climate Deal by "Early 2020".' Accessed September 03, 2019. https://www.climatechangenews.com/2019/06/14/eu-aims-long-term-climate-deal-early-2020/.

CNA. 2007. 'National Security and the Threat of Climate Change.' Accessed October 17, 2010. http://securityandclimate.cna.org/.

CNBC. 2019. 'EU's Juncker Proposes Billions of Spending on Climate Change After a 16-Year-Old's Speech.' Accessed February 25, 2019. https://www.cnbc.com/2019/02/22/juncker-proposes-climate-change-spending-after-greta-thunberg-speech.html.

Council of the EU. 2009. 'Council Conclusions on EU Position for the Copenhagen Climate Conference.' Accessed August 20, 2020. https://www.consilium.europa.eu/uedocs/cms_data/docs/pressdata/en/envir/110634.pdf.

Council of the European Union. 1993. *Council Decision Concerning the Conclusion of the United Nations Framework Convention on Climate Change.* 94/69/EC L 33/11. *Official Journal of the European Communities*, no. L33/11, 7.2.1994.

Council of the European Union. 1996a. 'Press Release. 1939th Council Meeting. Environment. Pollution Standards for the Engines of Non-Road Mobile Machinery.' 96/188. Accessed July 17, 2018. http://europa.eu/rapid/press-release_PRES-96-188_en.htm?locale=en.

Council of the European Union. 1996b. 'Press Release. 1939th Council Meeting. Environment. Pollution Standards for the Engines of Non-Road Mobile Machinery.' 96/188. Accessed July 17, 2018. http://europa.eu/rapid/press-release_PRES-96-188_en.htm?locale=en.

Council of the European Union. 2007a. 'Press Release. 2785th Council Meeting. Environment.' 6272/07. Accessed November 03, 2016. https://www.consilium.europa.eu/ueDocs/cms_Data/docs/pressData/en/envir/92864.pdf.

Council of the European Union. 2007b. 'Press Release. 2785th Council Meeting Environment: EU Objectives for the Further Development of the International Climate Regime Beyond 2012.' 6272/07.

Council of the European Union. 2010. *Presidency Conclusions: Climate Finance*, no. 16508/10. Brussels: Council of the European Union.

Council of the European Union. 2011. 'A Roadmap for Moving to a Competitive Low Carbon Economy in 2050 – Presidency Conclusions.' 11964/11.

Council of the European Union. 2012. 'A Roadmap for Moving to a Competitive Low-Carbon Economy in 2050- Presidency Conclusions.' 6842/12.

Council of the European Union. 2015. 'Press Release. EU Position for the UN Climate Change Conference in Paris: Council Conclusions.' 657/15.

Council of the European Union. 2018. 'Council Conclusions. Preparations for the UNFCCC Meetings in Katowice (2–14 December 2018).' 12901/18.

Council of the European Union. 2019. 'Climate Diplomacy. Council Conclusions.' 6153/19.

Council of the European Union. 2020. 'Council Conclusions on Climate Diplomacy.' 5033/20.

Council of the European Union, and European Commission. 1992. *Maastricht Treaty on European Union.* Luxembourg: Office for Official Publications of the European Communities.

Cox, Robert W. 1981. 'Social Forces, States and World Orders: Beyond International Relations Theory.' *Millennium* 10 (2): 126–55.

Cross, Mai'a K. D. 2017. 'The EU's Green Diplomacy.' Paper presented at the GLOBUS Conference on Conceptualising Global Justice, 19–20 January, Oslo.'.

Cross, Mai'a K. D. 2018. 'Partners at Paris? Climate Negotiations and Transatlantic Relations.' *Journal of European Integration* 40 (5): 571–86.

CSIS. 2017. 'The U.S.-China Climate and Energy Relationship.' Accessed July 10, 2020. https://www.csis.org/us-china-climate-and-energy-relationship.

Curtin, Joseph. 2010. *The Copenhagen Conference: How Should the EU Respond?* Dublin: Institute of International and European Affairs.

Dasgupta, Saibal. 2009. 'Copenhagen Conference: India, China Plan Joint Exit.' *Times of India*, October 28. Accessed November 24, 2020. https://timesofindia.indiatimes.com/india/Copenhagen-conference-India-China-plan-joint-exit/articleshow/5279771.cms.

Davis Cross, Mai'a K. 2017. 'The EU's Green Diplomacy.' Paper presented at the GLOBUS Conference on Conceptualising Global Justice, 19–20 January, Oslo.

Dehm, Julia. 2016. 'Indigenous Peoples and REDD+ Safeguards: Rights as Resistance or as Disciplinary Inclusion in the Green Economy?' *JHRE* 7 (2): 170–217.

Delbeke, Jos, Ger Klaasen, and Stefaan Vergote. 2015. 'Climate Related Energy Policies.' In *EU Climate Policy Explained*, edited by Jos Delbeke and Peter Vis, 52–78. London: Routledge.

Delbeke, Jos, and Peter Vis. 2015. 'EU Climate Leadership in a Rapidly Changing World.' In *EU Climate Policy Explained*, edited by Jos Delbeke and Peter Vis, 4–25. London: Routledge.

Depledge, Joanna. 2016. *The Organization of Global Negotiations: Constructing the Climate Change Regime.* Abingdon: Taylor and Francis.

Detraz, Nicole, and Michele M. Betsill. 2009. 'Climate Change and Environmental Security: For Whom the Discourse Shifts.' *International Studies Perspectives* 10 (3): 303–20.

Die Welt. 2020. '"Sonderbelastung der deutschen Wirtschaft" – CDU-Streit über Klimaziele.' *Die Welt*, April 11. Accessed July 29, 2020. https://www.welt.de/politik/deutschland/article207201765/Corona-Krise-CDU-Streit-um-Einhaltung-der-Klimaziele.html.

Diez, Thomas. 2001. 'Europe as a Discursive Battleground: Discourse Analysis and European Integration Studies.' *Cooperation and Conflict* 36 (1): 5–38.

Diez, Thomas. 2005. 'Constructing the Self and Changing Others: Reconsidering "Normative Power Europe".' *Millennium – Journal of International Studies* 33 (3): 613–36.

Diez, Thomas, Franziskus von Lucke, and Zehra Wellmann. 2016. *The Securitisation of Climate Change: Actors, Processes and Consequences.* London: Routledge.

Dimitrov, Radoslav S. 2016. 'The Paris Agreement on Climate Change: Behind Closed Doors.' *Global Environmental Politics* 16 (3): 1–11.

Donnelly, Jack. 1999. 'The Social Construction of Human Rights.' In *Human Rights in Global Politics*, edited by Timothy Dunne and Nicholas J. Wheeler, 71–102. Cambridge: Cambridge University Press.

Donnelly, Jack. 2007. 'The Relative Universality of Human Rights.' *Human Rights Quarterly* 29 (2): 281–306.

Douma, Wybe T. 2000. 'The Precautionary Principle in the European Union.' *Review of European Community and International Environmental Law* 9 (2): 132–43.

Dröge, Susanne. 2016. 'The Paris Agreement 2015: Turning Point for the International Climate Regime.' *SWP Research Paper* RP4.

Dröge, Susanne, and Oliver Geden. 2015. 'Die EU und das Pariser Klimaabkommen: Ambitionen, strategische Ziele und taktisches Vorgehen.' *SWP Aktuell* 42.

Dröge, Susanne, and Oliver Geden. 2016. ‚Nach dem Pariser Klimaabkommen: Neue Herausforderungen für eine Vorreiterrolle der EU.' *SWP Aktuell* 16.

Dröge, Susanne, and Vijeta Rattani. 2019. 'After the Katowice Climate Summit.' *SWP Comment*, no. 9: 1–7.

Dubash, Navroz K. 2012. 'Climate Change and Parliament: Excerpts from Lok Sabha and Rajya Sabha Debates in 2009.' In *Handbook of Climate Change and India: Development, Politics and Governance*, edited by Navroz K. Dubash, 230–45. Abingdon: Routledge.

Dubash, Navroz K., Radhika Khosla, Narasimha D. Rao, and Ankit Bhardwaj. 2018. 'India's Energy and Emissions Future: An Interpretive Analysis of Model Scenarios.' *Environmental Research Letters* 13 (7): 1–11.

DW. 2019. 'German Political Parties Go Green for EU "Climate Election".' Accessed July 29, 2020. https://www.dw.com/en/german-political-parties-go-green-for-eu-climate-election/a-48816129.

Dworkin, Ronald. 2011. *Justice for Hedgehogs.* Cambridge, MA. Belknap Press of Harvard University Press.

Eakin, Hallie, and Amy L. Luers. 2006. 'Assessing the Vulnerability of Social-Environmental Systems.' *Annual Review of Environment and Resources* 31 (1): 365–94.

Earth Negotiations Bulletin. 2007. 'Summary of the Thirteenth Conference of Parties to the UN Framework Convention on Climate Change and Third Meeting of Parties to the Kyoto Protocol: 3–15 December 2007.' Accessed August 20, 2020. https://enb.iisd.org/download/pdf/enb12354e.pdf.

ECDPM. 2014. 'Common but Differentiated Responsibilities (CBDR): The Challenge of Rejuvenating an Old-School Principle Ahead of the 2015 Climate Negotiations.' Accessed October 06, 2017. http://ecdpm.org/talking-points/common-differentiated-responsibilities-cbdr-challenge-rejuvenating-old-school-principle-ahead-2015-climate-negotiations/.

Eckersley, Robyn. 2004. *The Green State: Rethinking Democracy and Sovereignty.* Cambridge, MA: MIT Press.

Eckersley, Robyn. 2012. 'Moving Forward in the Climate Negotiations: Multilateralism or Minilateralism?' *Global Environmental Politics* 12 (2): 24–42.

EEA. 2010. 'Recession Accelerates the Decline in EU Greenhouse Gas Emissions.' Accessed November 07, 2018. https://www.eea.europa.eu/highlights/recession-accelerates-the-decline-in.

EEAS. 2016. 'Global Strategy to Steer EU External Action in an Increasingly Connected, Contested and Complex World.' Accessed November 27, 2020. https://op.europa.eu/en/publication-detail/-/publication/3eaae2cf-9ac5-11e6-868c-01aa75ed71a1.

EEAS, and European Commission. 2013. 'EU Climate Diplomacy for 2015 and Beyond: Reflection Paper.' Accessed November 24, 2020. https://ec.europa.eu/clima/sites/clima/files/international/negotiations/docs/eeas_26062013_en.pdf.

EESC. 2017. 'Just Transition to Low-Carbon Economy.' Accessed April 01, 2019. https://europa.eu/newsroom/events/just-transition-low-carbon-economy_en.

Eriksen, Erik O. 2016. 'Three Conceptions of Global Political Justice.' *GLOBUS Research Papers* 1/2016.

Eriksen, Erik O. 2019. 'Banishing Dominance in Europe: The Case for Regional Cosmopolitanism.' *European Journal of International Relations* 42 (1): 1–25.

Escobar, Arturo. 2018. *Designs for the Pluriverse: Radical Interdependence, Autonomy, and the Making of Worlds.* New Ecologies for the Twenty-First Century. Durham, NC: Duke University Press.

Eshelman, Robert S. 2014. 'The Danger of Fair and Balanced.' *Columbia Journalism Review*, May 1. Accessed October 20, 2016. http://www.cjr.org/essay/the_danger_of_fair_and_balance.php.

EU Council. 2009. 'Compilation of EU Statements for COP15; CMP5; AWG-KP 10; AWG-LCA 8.' 17733/09. Accessed November 27, 2020. https://data.consilium.europa.eu/doc/document/ST-17733-2009-INIT/en/pdf.

EurActiv. 2003. 'EU Builds "Green Diplomacy Network".' Accessed July 24, 2018. https://www.euractiv.com/section/sustainable-dev/news/eu-builds-green-diplomacy-network/.

European Commission. 1988. 'Communication to the Council: "The Greenhouse Effect and the Community": Commission Work Programme Concerning the Evaluation of Policy Options to Deal with the "Greenhouse Effect" and Draft Resolution on the Greenhouse Effect and the Community.' COM(88) 656 final.

European Commission. 1991a. 'Press Release. First Community Strategy to Limit Carbon Dioxide (CO2) Emissions and Improve Energy Efficiency.' P/92/29. Accessed November 21, 2016. https://ec.europa.eu/commission/presscorner/detail/en/P_92_29.

European Commission. 1991b. 'Communication from the Commission to the Council: A Community Strategy to Limit Carbon Dioxide Emissions and to Improve Energy Efficiency.' SEC(91) 1744 final.

European Commission. 1994. 'Growth, Competitiveness, Employment: The Challenges and Ways Forward into the 21st Century. White Paper.'Accessed November24,2020.https://op.europa.eu/en/publication-detail/-/publication/0d563bc1-f17e-48ab-bb2a-9dd9a31d5004.

European Commission. 1997a. 'Communication from the Commission: An Overall View of Energy Policy and Actions.' COM(97) 167 final.

European Commission. 1997b. 'Communication from the Commission on Environment and Employment (Building a Sustainable Europe).' COM(97) 592 final.

European Commission. 2005. 'Communication from the Commission to the Council, the European Parliament, the European Economic and Social Committee and the Committee of the Regions. Winning the Battle against Global Climate Change.' COM(2005) 35 final.

European Commission. 2008. 'Press Release. Boosting Growth and Jobs by Meeting our Climate Change Commitments.' IP/08/80.

European Commission. 2009a. '2020 Climate & Energy Package.' Accessed November 02, 2016. http://ec.europa.eu/clima/policies/strategies/2020/index_de.htm.

European Commission. 2009b. 'Communication from the Commission to the European Parliament, the Council, the European Economic and Social Committee and the Committee of the Regions: Towards a Comprehensive Climate Change Agreement in Copenhagen.' COM(2009) 39 final.

European Commission. 2010. 'Communication from the Commission to the European Parliament, the Council, the European Economic and Social Committee and the Committee of the Regions. Analysis of Options to Move Beyond 20%

Greenhouse Gas Emission Reductions and Assessing the Risk of Carbon Leakage.' COM(2010) 265 final.
European Commission. 2011. 'Communication from the Commission to the European Parliament, the Council, the European Economic and Social Committee and the Committee of the Regions: A Roadmap for Moving to a Competitive Low Carbon Economy in 2050.' COM(2011) 112 final.
European Commission. 2013a. 'Communication from the Commission to the European Parliament, the Council, the European Economic and Social Committee and the Committee of the Regions: The 2015 International Climate Change Agreement: Shaping International Climate Policy Beyond 2020.' COM(2013) 167 final.
European Commission. 2013b. 'Communication from the Commission to the European Parliament, the Council, the European Economic and Social Committee and the Committee of the Regions. An EU Strategy on Adaptation to Climate Change.' COM(2013) 216 final.
European Commission. 2013c. 'Press Release. Climate Action: Commission Proposes Ratification of Second Phase of Kyoto Protocol.' Accessed November 03, 2016. http://europa.eu/rapid/press-release_IP-13-1035_en.htm. IP/13/1035.
European Commission. 2014. 'Communication from the Commission to the European Parliament, the Council, the European Central Bank, the European Economic and Social Committee, the Committee of the Regions and the European Investment Bank. An Investment Plan for Europe.' COM(2014) 903 final.
European Commission. 2016a. 'Factsheet: The EU Emissions Trading System (EU ETS).' Accessed November 27, 2020. https://ec.europa.eu/clima/sites/clima/files/factsheet_ets_en.pdf.
European Commission. 2016b. 'Foreign Affairs Council Calls for Continuing European Climate Diplomacy Following Landmark Paris Deal.' Accessed November 24, 2020. https://ec.europa.eu/clima/news/articles/news_2016021601_en.
European Commission. 2016c. 'Press Release: Ministers Approve EU Ratification of Paris Agreement.' Accessed July 25, 2017. http://europa.eu/rapid/press-release_STATEMENT-16-3265_en.htm.
European Commission. 2018a. 'Paris Agreement.' Accessed July 25, 2018. https://ec.europa.eu/clima/policies/international/negotiations/paris_en.
European Commission. 2018b. 'Communication from the Commission to the European Parliament, the European Council, the Council, the European Economic and Social Committee, the Committee of the Regions and the European Investment Bank: A Clean Planet for All. A European Strategic Long-Term Vision for a Prosperous, Modern, Competitive and Climate Neutral Economy.' COM(2018) 773 final.
European Commission. 2019. 'The European Green Deal: Communication from the Commission to the European Parliament, the European Council, the Council, the European Economic and Social Committee and the Committee of the Regions.' COM(2019) 640 final.
European Council. 1990. 'Declaration by the European Council on the Environmental Imperative.' EC 6–1990.
European Council. 2003. 'A Secure Europe in a Better World: European Security Strategy.' Accessed November 24, 2020. https://www.consilium.europa.eu/en/documents-publications/publications/european-security-strategy-secure-europe-better-world/.

European Council. 2005. 'Brussels European Council. 22 and 23 March 2005: Presidency Conclusions.' 7619/1/05.
European Council. 2007a. 'Conclusions. Energy Policy for Europe: A contribution of the Energy Ministers to the 2007 Spring European Council.' 5411/07.
European Council. 2007b. 'Brussels European Council. 21 and 22 June 2007: Presidency Conclusions.' 11177/1/07.
European Council. 2007c. 'Brussels European Council 8 and 9 March 2007: Presidency Conclusions.' 7224/1/07.
European Council. 2009a. 'Brussels European Council 19 and 20 March 2009: Presidency Conclusions.' 7880/09.
European Council. 2009b. 'Brussels European Council 29 and 30 October 2009: Presidency Conclusions.' 15265/1/09 REV 1.
European Council. 2010a. 'European Council 25 and 26 March 2010: Conclusions.' EUCO 7/1/10 REV 1.
European Council. 2010b. 'European Council 28 and 29 October 2010: Conclusions.' EUCO 25/1/10 REV 1.
European Council. 2011. 'European Council 4 February 2011: Conclusions.' EUCO 2/1/11 REV 1.
European Council. 2014a. 'European Council: 20 and 21 March 2014: Conclusions.' EUCO 7/1/14 REV 1.
European Council. 2014b. 'European Council 26 and 27 June: Conclusions.' EUCO 79/14.
European Council. 2019a. 'European Council 17 and 18 October 2019: Conclusions.' EUCO 23/19.
European Council. 2019b. 'Speech by President Charles Michel at the UN Climate Conference COP25.' Accessed April 08, 2020. consilium.europa.eu/en/press/press-releases/2019/12/02/speech-by-president-charles-michel-at-the-un-climate-conference-cop25/.
European Council. 2019c. 'European Council 12 December 2019: Conclusions.' EUCO 29/19.
European Parliament. 1998a. 'Legislative Resolution Embodying Parliament's Opinion on the Proposal for a Council Directive on Transportable Pressure Equipment (COM(96)0674 – C4-0068/97 – 97/0011(SYN)).' *Official Journal of the European Communities* C 80/217.
European Parliament. 1998b. 'Resolution on Climate Change in the Run-Up to Buenos Aires (November 1998).' *Official Journal of the European Communities* B4-0802/98. C 313.
European Parliament. 2002. 'Outcome of the Bonn Conference on Climate Change. B5-0539, 0540, 0541, 0543, 0551 and 0552/2001. European Parliament Resolution on the Outcome of the Bonn Conference on Climate Change.' *Official Journal of the European Communities* C72E/321.
European Parliament, and Council of the European Union. 2003. 'Directive 2003/87/EC of the European Parliament and of the Council of 13 October 2003 Establishing a Scheme for Greenhouse Gas Emission Allowance Trading within the Community and Amending Council Directive 96/61/EC.' *Official Journal of the European Union* L 275/32.
European Parliament, and Council of the European Union. 2009. 'Decision No 406/2009/EC of the European Parliament and of the Council of 23 April 2009 on the Effort of Member States to Reduce their Greenhouse Gas Emissions to

Meet the Community's Greenhouse Gas Emission Reduction Commitments up to 2020.' *Official Journal of the European Union* L 140/136.

European Parliament, and Council of the European Union. 2013. 'Regulation (EU) No 525/2013 of the European Parliament and of the Council of 21 May 2013 on a Mechanism for Monitoring and Reporting Greenhouse Gas Emissions and for Reporting other Information at National and Union Level Relevant to Climate Change and Repealing Decision No 280/2004/EC.' *Official Journal of the European Union* L 165/13.

European Union. 2008. 'Report on the Implementation of the European Security Strategy – Providing Security in a Changing World.' S407/08. Accessed November 24, 2020. http://www.consilium.europa.eu/ueDocs/cms_Data/docs/pressdata/EN/reports/104630.pdf.

Falkner, Robert. 2007. 'The Political Economy of "Normative Power" Europe: EU Environmental Leadership in International Biotechnology Regulation.' *Journal of European Public Policy* 14 (4): 507–26.

Falkner, Robert. 2016. 'The Paris Agreement and the New Logic of International Climate Politics.' *International Affairs* 92 (5): 1107–25.

Falkner, Robert. 2018. 'International Climate Politics between Pluralism and Solidarism: An English School Perspective.' In *Traditions and Trends in Global Environmental Politics: International Relations and the Earth*, edited by Olaf Corry and Hayley Stevenson, 26–44. Abingdon: Routledge.

Falkner, Robert, and Barry Buzan. 2019. 'The Emergence of Environmental Stewardship as a Primary Institution of Global International Society.' *European Journal of International Relations* 25 (1): 131–55.

Falkner, Robert, Hannes Stephan, and John Vogler. 2010. 'International Climate Policy after Copenhagen: Towards a "Building Blocks" Approach.' *Global Policy* 1 (3): 252–62.

Findlay, Matthew, and Taylor Dimsdale. 2009. 'Climate Action in Major Emerging Economies: Climate Action in Major Emerging Economies Brazil, China, India, Mexico and South Africa.'. Accessed November 24, 2020. https://www.e3g.org/docs/Plus-Five-%28US%29-Updated-260109-TD.pdf.

Finley-Brook, Mary, and Curtis Thomas. 2011. 'Renewable Energy and Human Rights Violations: Illustrative Cases from Indigenous Territories in Panama.' *Annals of the Association of American Geographers* 101 (4): 863–72.

Fjellheim, Eva M., and Florian Carl. 2020. '"Green" Colonialism is Ruining Indigenous Lives in Norway: Norway Continues to Allow the Construction of Wind Farms on Saami Land, Which Threatens Their Herding Livelihoods.' Accessed August 03, 2020. https://www.aljazeera.com/indepth/opinion/colonialism-ruining-indigenous-lives-norway-200703135059280.html.

Floyd, Rita. 2011. 'Can Securitization Theory Be Used in Normative Analysis? Towards a Just Securitization Theory.' *Security Dialogue* 42 (4–5): 427–39.

Floyd, Rita. 2019a. 'Collective Securitisation in the EU: Normative Dimensions.' *West European Politics* 42 (2): 391–412.

Floyd, Rita. 2019b. *The Morality of Security: A Theory of Just Securitization.* Cambridge: Cambridge University Press.

Føllesdal, Andreas. 2000. 'Global Justice as Impartiality: Whither Claims to Equal Shares?' In *International Justice*, edited by A. J. Coates, 150–66. Avebury Series in Philosophy. Aldershot, Hampshire: Ashgate.

Fridays for Future. 2020. 'Unsere Forderungen an die Politik.' Accessed March 06, 2020. https://fridaysforfuture.de/forderungen/.

Friedman, Lisa. 2020. 'E.P.A., Citing Coronavirus, Drastically Relaxes Rules for Polluters.' *New York Times*, March 26. Accessed July 29, 2020. https://www.nytimes.com/2020/03/26/climate/epa-coronavirus-pollution-rules.html.

Füssel, Hans-Martin, and Richard J. T. Klein. 2006. 'Climate Change Vulnerability Assessments: An Evolution of Conceptual Thinking.' *Climatic Change* 75 (3): 301–29.

G77/China. 2008. 'Statement on Behalf of the Group of 77 and China by H.E. Amb Byron Blake, Deputy Permanent Representative of Antigua and Barbuda to the UN, at the 2nd Plenary of the 3rd Session of the AWG-LCA.' Accessed September 26, 2017. http://www.g77.org/statement/getstatement.php?id=080823.

Gajevic Sayegh, Alexandre. 2016. 'Justice in a Non-Ideal World: The Case of Climate Change.' *Critical Review of International Social and Political Philosophy* 28 (5): 1–26.

Gardiner, Stephen M. 2004a. 'Ethics and Global Climate Change.' *Ethics* 114 (3): 555–600.

Gardiner, Stephen M. 2004b. 'The Global Warming Tragedy and the Dangerous Illusion of the Kyoto Protocol.' *Ethics & International Affairs* 18 (1): 23–39.

Gardiner, Stephen M. 2006. 'A Perfect Moral Storm: Climate Change, Intergenerational Ethics and the Problem of Moral Corruption.' *Environmental Values* 15 (3): 397–413.

Gardiner, Stephen M., Simon Caney, Dale Jamieson, and Henry Shue, eds. 2010. *Climate Ethics: Essential Readings.* Oxford: Oxford University Press.

GCCA. 2015a. 'The Plus of GCCA+: The Global Climate Change Alliance Plus: An EU Flagship Initiative Supporting Climate Resilience.' Accessed November 24, 2020. http://www.gcca.eu/sites/default/files/gcca_concept_note.pdf.

GCCA. 2018a. 'Policy Dialogue.' Accessed November 08, 2018. http://www.gcca.eu/services/policy-dialogue.

GCCA+. 2015b. 'The Plus of GCCA+: The Global Climate Change Alliance Plus: An EU flagship Initiative Supporting Climate Resilience.' Accessed November 08, 2018. http://www.gcca.eu/sites/default/files/gcca_concept_note.pdf. Ref. Ares (2015) 5956167-18/12/2015.

GCCA+. 2018b. 'Building Climate Resilience: Mitigation and Adaptation in a Changing Climate: The Global Climate Change Alliance Plus (GCCA+).'. Accessed November 27, 2020. https://gcca.eu/sites/default/files/documents/2019-03/Building%20resilience%20to%20Climate%20Change_Blue_19032019_0.pdf

Geden, Oliver. 2013. 'Die Implementierung der „Kyoto -II" – Verpflichtungen in EU – Recht: Enger werdende Spielräume für eine klimapolitische Vorreiterrolle Deutschlands.' *SWP Aktuell* 69.

George, Alexander L., and Andrew Bennett. 2005. *Case Studies and Theory Development in the Social Sciences.* BCSIA Studies in International Security. Cambridge, MA: MIT Press.

Geyer, Florian, and Elspeth Guild. 2016. *Security Versus Justice? Police and Judicial Cooperation in the European Union.* London: Taylor and Francis.

Giroux, Henry A. 2006. 'Reading Hurricane Katrina: Race, Class, and the Biopolitics of Disposability.' *College Literature* 33 (3): 171–96.

Goldenberg, Suzanne. 2011. 'UN Security Council to Consider Climate Change Peacekeeping.' *The Guardian*, July 20. Accessed March 25, 2019. https://www.theguardian.com/environment/2011/jul/20/un-climate-change-peacekeeping.

Government of India. 2015. 'India's Intended Nationally Determined Contribution: Working towards Climate Justice.' Accessed August 20, 2020. https://www4.unfccc.int/sites/submissions/INDC/Published%20Documents/India/1/INDIA%20INDC%20TO%20UNFCCC.pdf.

Groen, Lisanne. 2016. 'The Importance of Fitting Activities to Context: The EU in Multilateral Climate and Biodiversity Negotiations.' PhD Thesis, Vrije Universiteit Brussels.

Groen, Lisanne, Arne Niemann, and Sebastian Oberthür. 2012. 'The EU as a Global Leader? The Copenhagen and Cancun UN Climate Change Negotiations.' *Journal of Contemporary European Research* 8 (2): 173–91.

Gupta, Joydeep, and Juhi Chaudhary. 2014. 'Walk the Talk on Climate, BASIC Group Tells Developed Countries.' *India Climate Dialogue*, August 8. Accessed August 20, 2020. https://indiaclimatedialogue.net/2014/08/08/walk-talk-climate-basic-group-tells-developed-countries/.

Hallding, Karl, Marie Ollson, Aaron Atteridge, Antto Vihma, Marcus Carson, and Mikael Román. 2011. *Together Alone: BASIC Countries and the Climate Change Conundrum.* Copenhagen: Nordic Council of Ministers.

Hansen, Lene. 2012. 'Reconstructing Desecuritisation: The Normative-Political in the Copenhagen School and Directions for How to Apply It.' *Review of International Studies* 38 (3): 525–46.

Harris, Paul G., ed. 2001. *The Environment, International Relations, and U.S. Foreign Policy.* Washington, DC: Georgetown University Press.

Harris, Paul G. 2002a. 'Sharing the Burdens of Environmental Change: Comparing EU and U.S. Policies.' *The Journal of Environment & Development* 11 (4): 380–401.

Harris, Paul G. 2002b. 'Environmental Security: Will Bush Follow Clinton's Lead?' *Pacifica Review: Peace, Security & Global Change* 14 (2): 149–57.

Harvey, Fiona. 2015. 'Paris Climate Change Agreement: The World's Greatest Diplomatic Success.' *The Guardian*, December 14. Accessed July 25, 2018. https://www.theguardian.com/environment/2015/dec/13/paris-climate-deal-cop-diplomacy-developing-united-nations.

Hein, Jonas. 2013. 'Climate Change Mitigation in Emerging Economies: The Case of Indonesia – Hot Air or Leadership?' d.i.e Briefing Paper 8/2013.

Heyward, Clare, and Dominic Roser, eds. 2016. *Climate Justice in a Non-Ideal World.* Oxford: Oxford University Press.

Hillebrand, Rainer. 2013. 'Climate Protection, Energy Security, and Germany's Policy of Ecological Modernisation.' *Environmental Politics* 22 (4): 664–82.

Hobson, John M. 2012. *The Eurocentric Conception of World Politics: Western International Theory, 1760–2010.* Cambridge: Cambridge University Press.

Hochstetler, Kathryn, and Manjana Milkoreit. 2014. 'Emerging Powers in the Climate Negotiations: Shifting Identity Conceptions.' *Political Research Quarterly* 67 (1): 224–35.

Hochstetler, Kathryn, and Manjana Milkoreit. 2015. 'Responsibilities in Transition: Emerging Powers in the Climate Change Negotiations.' *Global Governance* 21 (2): 205–26.

Hochstetler, Kathryn, and Eduardo Viola. 2012. 'Brazil and the Politics of Climate Change: Beyond the Global Commons.' *Environmental Politics* 21 (5): 753–71.

Holsti, Kalevi J. 2004. *Taming the Sovereigns: Institutional Change in International Politics.* Cambridge: Cambridge University Press.

Hovi, Jon, Tora Skodvin, and Stine Aakre. 2013. 'Can Climate Change Negotiations Succeed?' *Politics and Governance* 1 (2): 138–50.

Howard, Brian C. 2014. 'Brazil Leads World in Reducing Carbon Emissions by Slashing Deforestation.' *National Geographic*, June 5. Accessed August 20, 2020.

https://www.nationalgeographic.com/news/2014/6/140605-brazil-deforestation-carbon-emissions-environment/.
Huggan, Graham, and Helen Tiffin. 2007. 'Green Postcolonialism.' *Interventions* 9 (1): 1–11.
Hurrell, Andrew. 2007. *On Global Order: Power, Values, and the Constitution of International Society*. Oxford: Oxford University Press.
Hurrell, Andrew, and Sandeep Sengupta. 2012a. 'Emerging Powers, North-South Relations and Global Climate Politics.' *International Affairs* 88 (3): 463–84.
Hurrell, Andrew, and Sandeep Sengupta. 2012b. 'Emerging Powers, North-South Relations and Global Climate Politics.' *International Affairs* 88 (3): 463–84.
Hurst, Luke. 2020. 'Watch: Statues Under Attack as Europe Confronts Colonial Past.' *Euronews*. June 10. Accessed August 26, 2020. ttps://www.euronews.com/2020/06/10/watch-statues-under-attack-as-europe-confronts-colonial-past.
Ikeme, Jekwu. 2003. 'Equity, Environmental Justice and Sustainability: Incomplete Approaches in Climate Change Politics.' *Global Environmental Change* 13 (3): 195–206.
Ingram, James D. 2016. 'Cosmopolitanism from Below: Universalism as Contestation.' *Critical Horizons* 17 (1): 66–78.
International Solar Alliance. 2015. 'International Solar Alliance (ISA).' Accessed July 10, 2020. https://isolaralliance.org/.
Interview. 2013a. Interview by S. Aamodt. December 2. Brasilia. 02.12.2013.
Interview. 2013b. Interview by S. Aamodt. December 4. Brasilia. 04.12.2013.
Interview. 2013c. Interview by S. Aamodt. December 6. Brasilia. 06.12.2013.
Interview. 2014a. Interview by S. Aamodt. October 7. New Dehli. 07.10.2014.
Interview. 2014b. Interview by S. Aamodt. October 9. New Dehli. 09.10.2014.
Interview. 2014c. Interview by S. Aamodt. October 9. New Dehli. 09.10.2014.
Interview. 2014d. Interview by S. Aamodt. October 14. New Dehli. 14.10.2014.
Interview. 2014e. Interview by S. Aamodt. October 16. New Dehli. 16.10.2014.
Interview. 2017a. Interview by B. Ahrens. May 18. Brussels. 18.05.2017.
Interview. 2017b. Interview by B. Ahrens. May 18. Brussels. 18.05.2017.
Interview. 2017c. Interview by B. Ahrens. May 19. Brussels. 19.05.2017.
Interview. 2017d. Interview by B. Ahrens. May 23. Brussels. 23.05.2017.
Interview. 2017e. Interview by S. Aamodt. November 8. Bonn. 08.11.2017.
Interview. 2017f. Interview by S. Aamodt. November 8. Bonn. 08.11.2017.
Interview. 2017g. Interview by S. Aamodt. November 9. Bonn. 09.11.2017.
Interview. 2017h. Interview by S. Aamodt. November 9. Bonn. 09.11.2017.
Interview. 2018a. Interview by S. Aamodt. February 21. 21.02.2018.
Interview. 2018b. Interview by F. von Lucke. March 5. Brussels. 5.3.2018 (II).
Interview. 2018c. Interview by F. von Lucke. March 5. Brussels. 05.03.2018 (I).
Interview. 2018d. Interview by F. von Lucke. March 5. Brussels. 05.03.2018 (III).
Interview. 2018e. Interview by F. von Lucke. March 6. Brussels. 6.3.2018 (I).
Interview. 2018f. Interview by F. von Lucke. March 6. Brussels. 6.3.2018 (II).
Interview. 2018g. Interview by F. von Lucke. March 6. Brussels. 6.3.2018 (III).
Interview. 2018h. Interview by F. von Lucke. March 7. Brussels. 7.3.2018 (I).
Interview. 2018i. Interview by F. von Lucke. March 7. Brussels. 7.3.2018 (II).
Interview. 2018j. Interview by F. von Lucke. March 8. Brussels. 8.3.2018 (I).
Interview. 2018k. Interview by F. von Lucke. March 8. Brussels. 8.3.2018 (II).
Interview. 2018l. Interview by F. von Lucke. March 8. Brussels. 8.3.2018 (IIII).
Interview. 2018m. Interview by S. Aamodt. March 14. Brussels. 14.03.2018.
Interview. 2019a. Interview by S. Aamodt. February 26. 26.02.2019.

Interview. 2019b. Interview by S. Aamodt. February 26. Oslo. 26.02.2019.

Interview. 2019c. Interview by S. Aamodt. February 27. Oslo. 27.02.2019.

Interview. 2019d. Interview by S. Aamodt. March 11. 11.03.2019.

IPCC. 2010. 'The IPCC's Fifth Assessment Report (AR5).' Accessed November 15, 2010. http://www.ipcc.ch/pdf/press/ipcc_leaflets_2010/ipcc_ar5_leaflet.pdf.

IPCC. 2014. *Climate Change 2014 Synthesis Report: Contribution of Working Groups I, II and III to the Fifth Assessment Report of the Intergovernmental Panel on Climate Change*. Geneva: Intergovernmental Panel on Climate Change.

IPCC. 2015. *Climate Change 2014 Synthesis Report: Contribution of Working Groups I, II and III to the Fifth Assessment Report of the Intergovernmental Panel on Climate Change*. Geneva: Intergovernmental Panel on Climate Change.

IPCC. 2018. 'Summary for Policymakers of IPCC Special Report on Global Warming of 1.5°C Approved by Governments.' Accessed October 16, 2018. http://www.ipcc.ch/news_and_events/pr_181008_P48_spm.shtml.

Jobst, Conrad. 2010. ‚Sozialwissenschaftliche Analyse von Klimaforschung, -diskurs und -politik am Beispiel des IPCC.' In *Der Klimawandel*, edited by Martin Voss, 101–15. Wiesbaden: VS Verlag für Sozialwissenschaften.

Jong, Sijbren de, and Simon Schunz. 2012. 'Coherence in European Union External Policy Before and After the Lisbon Treaty: The Cases of Energy Security and Climate Change.' *European Foreign Affairs Review* 17 (2): 165–87.

Jordan, Andrew, and Tim Rayner. 2010. 'The Evolution of Climate Policy in the European Union: A Historical Overview.' In *Climate Change Policy in the European Union: Confronting the Dilemmas of Mitigation and Adaptation?*, edited by Andrew Jordan, Dave Huitema, Harro van Asselt, Tim Rayner, and Frans Berkhout, 52–80. Cambridge: Cambridge University Press.

Juncker, Jean-Claude. 2016. 'State of the Union Address 2016: Towards a Better Europe – A Europe that Protects, Empowers and Defends.' Accessed October 24, 2016. http://europa.eu/rapid/press-release_SPEECH-16-3043_en.htm.

Kane, John. 1996. 'Justice, Impartiality, and Equality: Why the Concept of Justice Does Not Presume Equality.' *Political Theory* 24 (3): 375–93.

Kant, Immanuel, and Hans S. Reiss. 1991. *Kant: Political Writings.* 2nd enlarged edition. Cambridge Texts in the History of Political Thought. Cambridge: Cambridge University Press.

Karstensen, Jonas, Glen P. Peters, and Robbie M. Andrew. 2018. 'Trends of the EU's Territorial and Consumption-Based Emissions from 1990 to 2016.' *Climatic Change* 151 (2): 131–42.

Kasa, Sjur, Anne T. Gullberg, and Gørild Heggelund. 2008. 'The Group of 77 in the International Climate Negotiations: Recent Developments and Future Directions.' *International Environmental Agreements: Politics, Law and Economics* 8 (2): 113–27.

Kelly, P. M., and Neil W. Adger. 2000. 'Theory and Practice in Assessing Vulnerability to Climate Change and Facilitating Adaptation.' *Climatic Change* 47 (4): 325–52.

Keohane, Robert O., and Michael Oppenheimer. 2016. 'Paris: Beyond the Climate Dead End through Pledge and Review?' *Politics and Governance* 4 (3): 142–51.

Keohane, Robert O., and David G. Victor. 2011. 'The Regime Complex for Climate Change.' *Perspectives on Politics* 9 (1): 7–23.

Kinley, Richard. 2016. 'Climate Change after Paris: From Turning Point to Transformation.' *Climate Policy* 17 (1): 9–15.

Kopra, Sanna. 2012. 'Is China a Responsible Developing Country? Climate Change Diplomacy and National Image Building.' *Global Asia Journal* (Occasional Paper No. 13).

Kraft, Michael E. 2013. 'Environmental Policy in Congress.' In *Environmental Policy: New Directions for the Twenty-First Century*, edited by Norman J. Vig and Michael E. Kraft, 8th ed., 109–34. Thousand Oaks, CA: CQ Press.

Kramer, Ronald C., and Raymond J. Michalowski. 2012. 'Is Global Warming a State-Corporate Crime?' In *Climate Change from a Criminological Perspective*, edited by Rob White, 71–88. New York: Springer.

La Rovere, Emilio L., Laura V. de Macedo, and Kevin A. Baumert. 2002. 'The Brazilian Proposal on Relative Responsibility for Global Warming.' In *Building on the Kyoto Protocol: Options for Protecting the Climate*, edited by Kevin A. Baumert, 157–74. Washington, DC: World Resources Institute.

Laatikainen, Katie V., and Karen E. Smith. 2006. *The European Union at the United Nations: Intersecting Multilateralism*. Basingstoke: Palgrave.

Laclau, Ernesto, and Chantal Mouffe. 1985. *Hegemony and Socialist Strategy: Towards a Radical Democratic Politics*. London: Verso.

Lahn, Bård, and Göran Sundqvist. 2017. 'Science as a "Fixed Point"? Quantification and Boundary Objects in International Climate Politics.' *Environmental Science & Policy* 67: 8–15.

Latham, Robert. 1997. *The Liberal Moment: Modernity, Security, and the Making of Postwar International Order*. New Directions in World Politics. New York: Columbia University Press. http://www.h-net.org/review/hrev-a0a1t7-aa.

Lauterpacht, Hersch. 1946. 'The Grotian Tradition in International Law.' *British Yearbook of International Law* 23: 1–53.

Lele, Sharachchandra. 2012. 'Climate Change and the Indian Environmental Movement.' In *Handbook of Climate Change and India: Development, Politics and Governance*, edited by Navroz K. Dubash, 208–17. Abingdon: Routledge.

Lenschow, Andrea, and Carina Sprungk. 2010. 'The Myth of a Green Europe.' *Journal of Common Market Studies* 48 (1): 133–54.

Li, Xinlei. 2017. 'China: From a Marginalized Follower to an Emerging Leader in Climate Politics.' In *The European Union in International Climate Change Politics: Still Taking a Lead?* edited by Rüdiger K. W. Wurzel, James Connelly, and Duncan Liefferink, 254–69. Abingdon: Routledge.

Lightfoot, Simon, and J. Burchell. 2005. 'The European Union and the World Summit on Sustainable Development: Normative Power Europe in Action?' *Journal of Common Market Studies* 43 (1): 75–95.

Linklater, Andrew, and Hidemi Suganami. 2006. *The English School of International Relations: A Contemporary Reassessment* 102. Cambridge: Cambridge University Press.

Little, Richard, and John Williams, eds. 2006. *The Anarchical Society in a Globalized World*. Basingstoke: Palgrave Macmillan.

Lockwood, Matthew. 2018. 'Right-Wing Populism and the Climate Change Agenda: Exploring the Linkages.' *Environmental Politics* 27 (4): 712–32.

López-Alves, Fernando. 2012. 'The Latin American Nation-State and the International.' In *Thinking International Relations Differently*, edited by Arlene B. Tickner and David L. Blaney, 161–80. London: Routledge.

Lovett, Frank. 2009. 'Domination and Distributive Justice.' *The Journal of Politics* 71 (3): 817–30.

Lucarelli, Sonia. 2018. 'The EU and the Crisis of Liberal Order: At Home and Abroad.' *GLOBUS Research Paper* 12/2018.

Lyons, Kristen, and Peter Westoby. 2014. 'Carbon Colonialism and the New Land Grab: Plantation Forestry in Uganda and Its Livelihood Impacts.' *Journal of Rural Studies* 36: 13–21.

Manners, Ian. 2002. 'Normative Europe: A Contradiction in Terms.' *Journal of Common Market Studies* 40 (2): 235–58.

Manners, Ian. 2006. 'Normative Power Europe Reconsidered: Beyond the Crossroads.' *Journal of European Public Policy* 13 (2): 182–99.

Markell, P. 2008. 'The Insufficiency of Non-Domination.' *Political Theory* 36 (1): 9–36.

Masquelier, Adeline. 2006. 'Why Katrina's Victims Aren't Refugees: Musings on a "Dirty" Word.' *American Anthropologist* 108 (4): 735–43.

Mathiesen, Karl, and Fiona Harvey. 2015. 'Climate Coalition Breaks Cover in Paris to Push for Binding and Ambitious Deal.' *The Guardian*, December 8. Accessed July 25, 2018. https://www.theguardian.com/environment/2015/dec/08/coalition-paris-push-for-binding-ambitious-climate-change-deal.

McCright, Aaron, and Riley E. Dunlap. 2011. 'The Politicization of Climate Change and Polarization in the American Public's View of Global Warming, 2001–2010.' *The Sociological Quarterly* 52: 155–94.

McDonald, Matt. 2011. 'Deliberation and Resecuritization: Australia, Asylum-Seekers and the Normative Limits of the Copenhagen School.' *Australian Journal of Political Science* 46 (2): 281–95.

McDonald, Matt. 2013. 'Discourses of Climate Security.' *Political Geography* 33 (March): 42–51.

Meadows, Damien, Yvon Slingenberg, and Peter Zapfel. 2015. 'EU ETS: Pricing Carbon to Drive Cost-Effective Reductions Across Europe.' In *EU Climate Policy Explained*, edited by Jos Delbeke and Peter Vis, 26–51. London: Routledge.

Methmann, Chris. 2010. '"Climate Protection" as Empty Signifier: A Discourse Theoretical Perspective on Climate Mainstreaming in World Politics.' *Millennium – Journal of International Studies* 39 (2): 345–72.

Meulemann, Max. 2015. 'The Purpose of INDCs for an International Climate Agreement.' *E-International Relations*, August 2015.

Meyer, Lukas H., and Dominic Roser. 2010. 'Climate Justice and Historical Emissions.' *Critical Review of International Social and Political Philosophy* 13 (1): 229–53.

Meyer, Lukas H., and Dominic Roser. 2013. 'Climate Justice and Historical Emissions.' In *Democracy, Equality, and Justice*, edited by Matt Matravers and Lukas H. Meyer, 229–54. Hoboken, NJ: Taylor and Francis.

Michaelowa, Axel, and Katharina Michaelowa. 2015. 'Do Rapidly Developing Countries Take Up New Responsibilities for Climate Change Mitigation?' *Climatic Change* 133 (3): 499–510.

Mildner, Stormy-Annika, and Jörn Richert. 2010. 'Obamas neue Klimapolitik: Möglichkeiten und Grenzen eines klimapolitischen Wandels in den USA.' *SWP Studie* February 2010. Accessed March 01, 2013. http://www.swp-berlin.org/fileadmin/contents/products/studien/2010_S04_mdn_rrt_ks.pdf.

Miller, Clark A. 2004. 'Climate Science and the Making of a Global Political Order.' In *States of Knowledge: The Co-Production of Science and Social Order*, edited by Sheila Jasanoff, 46–66. International Library of Sociology. London: Routledge.

Moellendorf, Darrel. 2015. 'Climate Change Justice.' *Philosophy Compass* 10 (3): 173–86.

Monheim, Kai. 2015. *How Effective Negotiation Management Promotes Multilateral Cooperation: The Power of Process in Climate, Trade, and Biosafety Negotiations.* Routledge Research in Global Environmental Governance, 2. London: Routledge.

Morales, David. 2015. 'Achievements and Challenges of the Paris Agreement.' Accessed October 10, 2017. https://anglicanalliance.org/achievements-and-challenges-of-the-paris-agreement/.

Moser, Susanne C. 2012. 'Adaptation, Mitigation, and Their Disharmonious Discontents: An Essay.' *Climatic Change* 111 (2): 165–75.

Müller-Rommel, Ferdinand. 2002. 'The Lifespan and the Political Performance of Green Parties in Western Europe.' *Environmental Politics* 11 (1): 1–16.

Najam, Adil. 2005. 'Developing Countries and Global Environmental Governance: From Contestation to Participation to Engagement.' *International Environmental Agreements: Politics, Law and Economics* 5 (3): 303–21.

Najam, Adil, Saleemul Huq, and Youba Sokona. 2003. 'Climate Negotiations Beyond Kyoto: Developing Countries Concerns and Interests.' *Climate Policy* 3 (3): 221–31.

Narlikar, Amrita, and Diana Tussie. 2004. 'The G20 at the Cancun Ministerial: Developing Countries and Their Evolving Coalitions in the WTO.' *World Economy* 27 (7): 947–66.

Nasiritousi, Naghmeh, Mattias Hjerpe, and Björn-Ola Linnér. 2016. 'The Roles of Non-State Actors in Climate Change Governance: Understanding Agency through Governance Profiles.' *International Environmental Agreements: Politics, Law and Economics* 16 (1): 109–26.

Neal, Andrew W. 2011. *Exceptionalism and the Politics of Counter-Terrorism: Liberty, Security and the War on Terror.* London: Routledge.

Nentjes, Andries, and Ger Klaassen. 2004. 'On the Quality of Compliance Mechanisms in the Kyoto Protocol.' *Energy Policy* 32 (4): 531–44.

Neslen, Arthur. 2015. 'EU to Launch Diplomatic Offensive ahead of Paris Climate Talks.' *The Guardian*, January 20. Accessed November 24, 2020. https://www.theguardian.com/environment/2015/jan/20/eu-to-launch-diplomatic-offensive-ahead-paris-climate-talks.

Neumayer, Eric. 2000. 'In Defence of Historical Accountability for Greenhouse Gas Emissions.' *Ecological Economics* 33 (2): 185–92.

Ngcoya, Mvuselelo. 2015. 'Ubuntu: Toward an Emancipatory Cosmopolitanism?' *International Political Sociology* 9 (3): 248–62.

Nicholls, Robert J., and Anny Cazenave. 2010. 'Sea-Level Rise and Its Impact on Coastal Zones.' *Science* 328 (5985): 1517–20.

Nye, Joseph S. 2004. *Soft Power: The Means to Success in World Politics.* 1st ed. New York: Public Affairs.

Obergassel, Wolfgang, Christof Arens, Lukas Hermwille, Nico Kreibich, Florian Mersmann, Hermann E. Ott, and Hanna Wang-Helmreich. 2016. 'Phoenix from the Ashes: An Analysis of the Paris Agreement to the UNFCCC.' https://epub.wupperinst.org/frontdoor/deliver/index/docId/6374/file/6374_Obergassel.pdf. Accessed July 13, 2017.

Oberthür, Sebastian. 2011. 'The European Union's Performance in the International Climate Change Regime.' *Journal of European Integration* 33 (6): 667–82.

Oberthür, Sebastian. 2016. 'Where to Go from Paris? The European Union in Climate Geopolitics.' *Global Affairs* 2 (2): 119–30.

Oberthür, Sebastian, and Claire Dupont. 2011. 'The Council, the European Council and International Climate Policy: From Symbolic Leadership to Leadership by Example.' In *The European Union as a Leader in International Climate Change Politics*, edited by Rüdiger Wurzel and James Connelly, 74–91. Routledge/UACES Contemporary European Studies 15. London: Routledge.

Oberthür, Sebastian, and Lisanne Groen. 2017. 'Explaining Goal Achievement in International Negotiations: The EU and the Paris Agreement on Climate Change.' *Journal of European Public Policy* 6 (1): 1–20.

Oberthür, Sebastian, and H. E. Ott. 1999. *The Kyoto Protocol: International Climate Policy for the 21st Century.* Berlin: Springer.

Oberthür, Sebastian, and Claire Roche Kelly. 2008. 'EU Leadership in International Climate Policy: Achievements and Challenges.' *The International Spectator* 43 (3): 35–50.

O'Brien, Karen, Siri Eriksen, Lynn P. Nygaard, and Ane Schjolden. 2007. 'Why Different Interpretations of Vulnerability Matter in Climate Change Discourses.' *Climate Policy* 7 (1): 73–88.

Oels, Angela, and Franziskus von Lucke. 2015. ‚Gescheiterte Versicherheitlichung oder Sicherheit im Wandel: Hilft uns die Kopenhagener Schule beim Thema Klimawandel?' *Zeitschrift für Internationale Beziehungen* 22 (1): 43–70.

OHCHR. 2015. 'Understanding Human Rights and Climate Change: Submission of OHCHR to the 21st Conference of the Parties to the UNFCCC.' Accessed November 24, 2020. http://www.ohchr.org/Documents/Issues/ClimateChange/COP21.pdf.

Ojha, Vijay P. 2009. 'Carbon Emissions Reduction Strategies and Poverty Alleviation in India.' *Environment and Development Economies* 14 (3): 323–48.

Okereke, Chukwumerije, Harriet Bulkeley, and Heike Schroeder. 2009. 'Conceptualizing Climate Governance beyond the International Regime.' *Global Environmental Politics* 9 (1): 58–78.

Okereke, Chukwumerije, and Philip Coventry. 2016. 'Climate Justice and the International Regime: Before, During, and After Paris.' *Wiley Interdisciplinary Reviews: Climate Change* 7 (6): 834–51.

Olivier, Jos G. J., Greet Janssens-Manhout, Marilena Muntean, and Jeroen A. H. W. Peters. 2015. 'Trends in Global CO2 Emissions 2015 Report'.

Oreskes, Naomi. 2004. 'The Scientific Consensus on Climate Change.' *Science* 306 (5702): 1686.

Ostrom, Elinor. 2010. 'A Polycentric Approach for Coping with Climate Change: Background Paper to the 2010 World Development Report.' *Policy Research Working Paper* 5095.

Page, Edward. 1999. 'Intergenerational Justice and Climate Change.' *Political Studies* 47 (1): 53–66.

Page, Edward. 2013. 'Climate Change Justice.' In *The Handbook of Global Climate Change and Environmental Policy*, edited by Robert Falkner, 231–47. Handbooks of Global Policy Series. Hoboken, NJ: Wiley-Blackwell.

Paltsev, Sergey, Jennifer Morris, Yongxia Cai, Valerie Karplus, and Henry Jacoby. 2012. 'The Role of China in Mitigating Climate Change.' *Energy Economics* 34 (Supplement 3): 444–50.

Papa, Mihaela, and Nancy W. Gleason. 2012. 'Major Emerging Powers in Sustainable Development Diplomacy: Assessing their Leadership Potential.' *Global Environmental Change* 22 (4): 915–24.

Parry, Martin L., Osvaldo F. Canziani, Jean P. Palutikof, Paul J. van der Linden, and Clair E. Hanson, eds. 2007. *Climate Change 2007: Impacts, Adaptation and Vulnerability. Contribution of Working Group II to the Fourth Assessment Report of the Intergovernmental Panel on Climate Change.* Cambridge: Cambridge University Press. Accessed March 01, 2013. http://www.ipcc.ch/pdf/assessment-report/ar4/wg2/ar4_wg2_full_report.pdf.

Pasternack, A. 2009. 'Tuvalu Negotiator Delivers a Tear-Jerking Call for the Toughest Climate Treaty at Copenhagen.' *Treehugger*, December 12.

Paterson, Matthew, and Johannes Stripple. 2010. 'My Space: Governing Individuals' Carbon Emissions.' *Environment and Planning D: Society and Space* 28 (2): 341–62.

Pauw, Pieter, Steffen Bauer, Carmen Richerzhagen, Clara Brandi, and Hanna Schmole. 2014. 'Different Perspectives on Differentiated Responsibilities: A State-of-the-Art Review of the Notion of Common but Differentiated Responsibilities in International Negotiations.' *DIE Discussion Paper* 6.

Pazzanese, Christina. 2016. 'Politics in a "Post-Truth" Age: In this Topsy-Turvy Presidential Campaign, the Old Laws May No Longer Apply.' *Harvardgazette*, July 14. Accessed October 20, 2016. http://news.harvard.edu/gazette/story/2016/07/politics-in-a-post-truth-age/.

Pettit, Philip. 1997. *Republicanism: A Theory of Freedom and Government.* Oxford: Oxford University Press.

Pettit, Philip. 2010. 'Legitimate International Institutions: A Neorepublican Perspective.' In *The Philosophy of International Law*, edited by Samantha Besson and John Tasioulas, 139–62. Oxford: Oxford University Press.

Pickering, Jonathan, Karin Bäckstrand, and David Schlosberg. 2020. 'Between Environmental and Ecological Democracy: Theory and Practice at the Democracy-Environment Nexus.' *Journal of Environmental Policy & Planning* 22 (1): 1–15.

Pickett, Brent L. 1996. 'Foucault and the Politics of Resistance.' *Polity* 28 (4): 445–66.

Poguntke, Thomas. 1987. 'New Politics and Party Systems: The Emergence of a New Type of Party?' *West European Politics* 10 (1): 76–88.

Posner, Eric A., David A. Weisbach, and Cass R. Sunstein. 2008. 'Climate Change Justice.' *Georgetown Law Journal* 96: 1565–612.

Prange, Astrid. 2019. 'Welche Zukunft hat Fridays for Future?' *Deutsche Welle*, August 4. Accessed September 02, 2020. https://www.dw.com/de/welche-zukunft-hat-fridays-for-future/a-49888377.

Qi, Xinran. 2011. 'The Rise of BASIC in UN Climate Change Negotiations.' *South African Journal of International Affairs* 18 (3): 295–318.

Rajamani, Lavanya. 2016. 'The 2015 Paris Agreement: Interplay Between Hard, Soft and Non-Obligations.' *Journal of Environmental Law* 28 (2): 337–58.

Rajamani, Lavanya. 2017. 'India's Approach to International Law in the Climate Change Regime.' *Indian Journal of International Law* 57 (1–2): 1–23.

Raman, Meenakshi. n.d. 'The Climate Change Battle in Paris: An Updated Analysis of the Paris COP21 and the Paris Agreement.' *Third World Network Briefing Paper.*

Randerson, James, Damian Carrington, and Adam Vaughan. 2010. 'Cancún Climate Change Summit: The Final Day as It Happened.' *The Guardian*, December 10.

Accessed July 25, 2018. https://www.theguardian.com/environment/blog/2010/dec/10/cancun-climate-change-summit-final.

Rawls, John. 2003. *A Theory of Justice*. Cambridge, MA: Belknap Press of Harvard University Press.

Rayner, Tim, and Andrew Jordan. 2016. 'Climate Change Policy in the European Union.' In *Oxford Research Encyclopedia of Climate Science*, edited by Tim Rayner and Andrew Jordan, 1–30: Oxford: Oxford University Press.

Reus-Smit, Christian. 2001. 'Human Rights and the Social Construction of Sovereignty.' *Review of International Studies* 27 (4): 519–38.

Richmond, Oliver P. 2017. 'The Paradox of Peace and Power: Contamination or Enablement?' *International Politics* 54 (5): 637–58.

Risse, Thomas, Steve C. Ropp, and Kathryn Sikkink, eds. 1999. *The Power of Human Rights: International Norms and Domestics Change*. New York: Cambridge University Press.

Roberts, J. T., and Bradley C. Parks. 2007. *A Climate of Injustice: Global Inequality, North-South Politics, and Climate Policy*. Global Environmental Accord. Cambridge, MA: MIT Press.

Roeck, Frederik de. 2019. 'Governmentality and the Climate-Development Nexus: The Case of the EU Global Climate Change Alliance.' *Global Environmental Change* 55: 160–67.

Roeck, Frederik de, Sarah Delputte, and Jan Orbie. 2016. 'Framing the Climate-Development Nexus in the European Union.' *Third World Thematics: A Third World Quarterly Journal* 1 (4): 437–53.

Rose, Adam. 1990. 'Reducing Conflict in Global Warming Policy: The Potential of Equity as a Unifying Principle.' *Energy Policy* 18 (10): 927–35.

Sachs, Benjamin. 2014. 'The Relevance of Distributive Justice to International Climate Change Policy.' *Ethics, Policy & Environment* 17 (2): 208–24.

Saran, Shyam. 2015. 'Paris Climate Talks: Developed Countries Must Do More than Reduce Emissions.' *The Guardian*, November 23. Accessed August 20, 2020. https://www.theguardian.com/environment/2015/nov/23/paris-climate-talks-developed-countries-must-do-more-than-reduce-emissions.

Scheipers, Sibylle, and Daniela Sicurelli. 2007. 'Normative Power Europe: A Credible Utopia?' *Journal of Common Market Studies* 45 (2): 435–57.

Schmidt, Susanne K. 2007. 'Mutual Recognition as a New Mode of Governance.' *Journal of European Public Policy* 14 (5): 667–81.

Schneider, Michael. 2017. 'An Assessment of the Past Performance and the Future Role of the High Ambition Coalition.' ICCG Reflection No. 52/April 2017.

Schreurs, Miranda A. 2002. *Environmental Politics in Japan, Germany, and the United States*. Cambridge: Cambridge University Press.

Schreurs, Miranda A., and Yves Tiberghien. 2007. 'Multi-Level Reinforcement: Explaining European Union Leadership in Climate Change Mitigation.' *Global Environmental Politics* 7 (4): 19–46.

Schroeder, Heike. 2010. 'Agency in International Climate Negotiations: The Case of Indigenous Peoples and Avoided Deforestation.' *International Environmental Agreements: Politics, Law and Economics* 10 (4): 317–32.

Schroeder, Heike, Maxwell T. Boykoff, and Laura Spiers. 2012. 'Equity and State Representations in Climate Negotiations.' *Nature Climate Change* 2: 834–36.

Schroeder, Heike, and Heather Lovell. 2012. 'The Role of Non-Nation-State Actors and Side Events in the International Climate Negotiations.' *Climate Policy* 12 (1): 23–37.

Schunz, Simon. 2012. 'Explaining the Evolution of European Union Foreign Climate Policy: A Case of Bounded Adaptiveness.' *European Integration online Papers (EIoP)* 16 (Article 6): 1–30.

Schunz, Simon. 2015. 'The European Union's Climate Change Diplomacy.' In *The European Union as a Diplomatic Actor*, edited by Joachim A. Koops and Gjovalin Macaj, 178–201. The European Union in International Affairs. Basingstoke: Palgrave Macmillan.

Schüssler, Rudolf. 2011. 'Climate Justice: A Question of Historic Responsibility?' *Journal of Global Ethics* 7 (3): 261–78.

Seck, Sara L. 2017. 'Revisiting Transnational Corporations and Extractive Industries: Climate Justice, Feminism, and State Sovereignty.' *Transnational Law and Contemporary Problems* 26 (2): 383–414.

Shapiro, Ian. 2012. 'On Non-Domination.' *University of Toronto Law Journal* 62 (3): 293–336.

Shue, Henry. 1993. 'Subsistence Emissions and Luxury Emissions.' *Law & Policy* 15 (1): 39–59.

Shue, Henry. 1999. 'Global Environment and International Inequality.' *International Affairs* 75 (3): 531–45.

Shue, Henry. 2010. 'Subsistence Emissions and Luxury Emissions.' In *Climate Ethics: Essential Readings*, edited by Stephen M. Gardiner, Simon Caney, Dale Jamieson, and Henry Shue, 200–14. Oxford: Oxford University Press.

Sierp, Aline. 2020. 'EU Memory Politics and Europe's Forgotten Colonial Past.' *Interventions* 22 (6): 686–702.

Sjursen, Helene. 2006a. 'The EU as a "Normative" Power: How Can this Be?' *Journal of European Public Policy* 13 (2): 235–51.

Sjursen, Helene. 2006b. 'What Kind of Power?' *Journal of European Public Policy* 13 (2): 169–81.

Sjursen, Helene. 2017. 'Global Justice and Foreign Policy: The Case of the European Union.' *GLOBUS Research Papers* 2/2017.

Skinner, Quentin. 2010. 'On the Slogans of Republican Political Theory.' *European Journal of Political Theory* 9 (1): 95–102.

Skovgaard, Jakob. 2014. 'EU Climate Policy after the Crisis.' *Environmental Politics* 23 (1): 1–17.

Smith, Samantha. 2017. 'Just Transition: A Report for the OECD.' Accessed November 27, 2020. https://extractiveshub.org/servefile/getFile/id/7453.

Solana, Javier, and European Commission. 2008. 'Climate Change and International Security. Paper from the High Representative and the European Commission to the European Council (S113/08).' Accessed September 21, 2010. http://www.consilium.europa.eu/uedocs/cms_data/docs/pressdata/en/reports/99387.pdf.

Soto, Àlvaro de. 2014. 'Diplomacy and Mediation.' In *The Handbook of Global Security Policy*. Vol. 16, edited by Iavor Rangelov and Mary Kaldor, 300–19. Handbook of Global Policy Series. Chichester: John Wiley & Sons Inc.

Springer, Allen L. 2008. 'The Failures of American and European Climate Policy: International Norms, Domestic Politics, and Unachievable Commitments – By Loren R. Cass.' *Review of Policy Research* 25 (5): 491–92.

Steinberg, Paul F. 2012. 'Welcome to the Jungle: Policy Theory and Political Instability.' In *Comparative Environmental Politics: Theory, Practice, and Prospects*, edited by Paul F. Steinberg and Stacy D. VanDeveer, 255–84. American and Comparative Environmental Policy. Cambridge, MA: MIT Press.

Stern, Nicholas. 2006. 'Stern Review on the Economics of Climate Change.' Accessed October 17, 2010. http://siteresources.worldbank.org/INTINDONESIA/Resources/226271-1170911056314/3428109-1174614780539/SternReviewEng.pdf.

Storey, David E. 2019. 'The World as it Is: A Vision for a Social Science (and Policy) Turn in Climate Justice.' In *A Research Agenda for Climate Justice*, edited by Paul G. Harris, 38–51. Elgar Research Agendas. Cheltenham: Edward Elgar.

Stuenkel, Oliver. 2016. 'Do the BRICS Possess Soft Power?' *Journal of Political Power* 9 (3): 353–67.

Suiseeya, Kimberly R. M. 2016. 'Transforming Justice in REDD+ through a Politics of Difference Approach.' *Forests* 7 (300): 1–14.

Tabau, A.-S., and S. Maljean-Dubois. 2010. 'Non-Compliance Mechanisms: Interaction between the Kyoto Protocol System and the European Union.' *European Journal of International Law* 21 (3): 749–63.

Thew, Harriet, Lucie Middlemiss, and Jouni Paavola. 2020. '"Youth Is Not a Political Position": Exploring Justice Claims-Making in the UN Climate Change Negotiations.' *Global Environmental Change* 61: 102036.

Tickner, Arlene B. 2003. 'Seeing IR Differently: Notes from the Third World.' *Millennium – Journal of International Studies* 32 (2): 295–324.

Tickner, Arlene B. 2013. 'Core, Periphery and (Neo)Imperialist International Relations.' *European Journal of International Relations* 19 (3): 627–46.

Torney, Diarmuid. 2015a. *European Climate Leadership in Question: Policies toward China and India*. Earth System Governance. Cambridge, MA: MIT Press.

Torney, Diarmuid. 2015b. 'The Role of the European External Action Service in Climate Diplomacy.' *GovInn Policy Brief* 8.

Torney, Diarmuid, and Mai'a K. D. Cross. 2018. 'Environmental and Climate Diplomacy: Building Coalitions Through Persuasion.' In *European Union External Environmental Policy: Rules, Regulation and Governance Beyond Borders*, edited by Camilla Adelle, Katja Biedenkopf, and Diarmuid Torney, 39–58. The European Union in International Affairs. Cham: Palgrave Macmillan.

Trombetta, Maria J. 2008. 'Environmental Security and Climate Change: Analysing the Discourse.' *Cambridge Review of International Affairs* 21 (4): 585–602.

Trump, Donald J. 2017. 'Statement by President Trump on the Paris Climate Accord.' Accessed February 16, 2020. https://www.whitehouse.gov/briefings-statements/statement-president-trump-paris-climate-accord/.

Tsosie, Rebecca A. 2007. 'Indigenous People and Environmental Justice: The Impact of Climate Change.' *University of Colorado Law Review* 78 (4): 1625–77.

Tyler, Emily. 2010. 'Aligning South African Energy and Climate Change Mitigation Policy.' *Climate Policy* 10 (5): 575–88.

UN. 1992. 'United Nations Framework Convention on Climate Change.' Accessed September 09, 2020. https://unfccc.int/resource/docs/convkp/conveng.pdf.

UN. 2015. 'Paris Agreement.' Accessed July 14, 2016. http://unfccc.int/files/essential_background/convention/application/pdf/english_paris_agreement.pdf.

UNDP. 2009. *Human Development Report 2009: Overcoming Barriers: Human Mobility and Development*. Human Development Report 2009. Basingstoke: Palgrave Macmillan.

UNESCO. 2005. 'The Precautionary Principle.' Accessed November 27, 2020. https://unesdoc.unesco.org/ark:/48223/pf0000139578.

UNFCCC. n.d. 'What is the AWG-KP?' Accessed November 27, 2020. https://unfccc.int/process-and-meetings/bodies/bodies-that-have-concluded-work/

ad-hoc-working-group-on-further-commitments-for-annex-i-parties-under-the-kyoto-protocol-awg-kp.

UNFCCC. 2007. 'Bali Action Plan.' decision 1/CP.13. Accessed November 24, 2020. http://unfccc.int/resource/docs/2007/cop13/eng/06a01.pdf.

UNFCCC. 2009a. 'Ideas and Proposals on the Elements Contained in Paragraph 1 of the Bali Action Plan Submissions from Parties Part II.' Accessed March 28, 2013. http://unfccc.int/resource/docs/2009/awglca6/eng/misc04p02.pdf.

UNFCCC. 2009b. 'Copenhagen Accord.' Accessed September 09, 2020. https://unfccc.int/resource/docs/2009/cop15/eng/11a01.pdf.

UNFCCC. 2012. 'Doha Amendment to the Kyoto Protocol.' Accessed November 03, 2016. http://unfccc.int/files/kyoto_protocol/application/pdf/kp_doha_amendment_english.pdf.

UNFCCC. 2013. 'Status of Ratification of the Kyoto Protocol.' Accessed March 28, 2013. http://unfccc.int/kyoto_protocol/status_of_ratification/items/2613.php.

UNFCCC. 2015. 'Synthesis Report on the Aggregate Effect of the Intended Nationally Determined Contributions.' Accessed July 26, 2018. https://unfccc.int/resource/docs/2015/cop21/eng/07.pdf.

UNFCCC. 2017. 'NDC Registry.' Accessed October 12, 2017. http://www4.unfccc.int/ndcregistry/Pages/Home.aspx.

UNFCCC. 2019a. 'Talanoa Call for Action.' Accessed September 03, 2019. https://unfccc.int/sites/default/files/resource/Talanoa%20Call%20for%20Action.pdf.

UNFCCC. 2019b. 'The Katowice Rulebook – Main Principles of the Document.' Accessed September 03, 2019. https://cop24.gov.pl/news/news-details/news/the-katowice-rulebook-main-principles-of-the-document/.

UNSC. 2007. '5663 Meeting, Tuesday, 17 April 2007, 3 p.m., S/PV5663 (Resumption 1)'.

UNSC. 2011. '6587th Meeting, Wednesday, 20 July 2011, 10 a.m.' Accessed February 28, 2013. http://www.circleofblue.org/waternews/wp-content/uploads/2012/09/UN-Security-Council_climate-change-July-2011.pdf.

UNSC. 2019. 'Climate Change Recognized as "Threat Multiplier"', UN Security Council Debates Its Impact on Peace.' Accessed March 21, 2019. https://news.un.org/en/story/2019/01/1031322.

Upadhyaya, Prabhat, Mathias Fridahl, Björn-Ola Linnér, and Mikael Román. 2018. 'Comparing Climate Policy Processes in India, Brazil, and South Africa: Domestic Engagements with International Climate Policy Frameworks.' *The Journal of Environment & Development* 27 (2): 186–209.

U.S. Government. 2006. *Hurricane Katrina: A Nation Still Unprepared: Special Report of the Committee on Homeland Security and Governmental Affairs, United States Senate, Together with Additional Views.* Report/109th Congress, 2d session, Senate 109–322. Washington, DC: U.S. Government Printing Office.

van der Heijden, Hein-Anton. 2010. *Social Movements, Public Spheres and the European Politics of the Environment: Green Power Europe?* Basingstoke: Palgrave Macmillan.

van Schaik, Louise G., and Simon Schunz. 2012. 'Explaining EU Activism and Impact in Global Climate Politics: Is the Union a Norm- or Interest-Driven Actor?' *Journal of Common Market Studies* 50 (1): 169–86.

van Tilburg, Xander, James Rawlins, and Shikha Bhasin. 2016. 'NAMA Development and Stakeholder Engagement: ECN Policy Studies.' Accessed November 24, 2020. https://www.transparency-partnership.net/sites/default/files/vantilburg_and_rawlins_2016__nama_development_and_stakeholder_engagement.pdf.

Victor, David G. 2015. *The Case for Climate Clubs.* Geneva: International Centre for Trade and Sustainable Development (ICTSD) and World Economic Forum. E15Initiative.

Vidal, John. 2009. 'Copenhagen Climate Summit in Disarray After "Danish Text" Leak.' *The Guardian*, December 8. Accessed August 20, 2020. https://www.theguardian.com/environment/2009/dec/08/copenhagen-climate-summit-disarray-danish-text.

Vidal, John. 2010. 'Copenhagen Climate Failure Blamed on "Danish Text".' *The Guardian*, May 31. Accessed August 20, 2020. https://www.theguardian.com/environment/2010/may/31/climate-change-copenhagen-danish-text.

Vig, Norman J. 2013. 'Presidential Powers and Environmental Policy.' In *Environmental Policy: New Directions for the Twenty-First Century,* edited by Norman J. Vig and Michael E. Kraft, 8th ed., 84–108. Thousand Oaks, CA: CQ Press.

Vig, Norman J., and Michael E. Kraft, eds. 2013. *Environmental Policy: New Directions for the Twenty-First Century.* 8th ed. Thousand Oaks, CA: CQ Press.

Vincent, R. J. 1986. *Human Rights and International Relations.* Cambridge: Cambridge University Press.

Vogler, John. 2005. 'The European Contribution to Global Environmental Governance.' *International Affairs* 81 (4): 835–50.

Voigt, Christina. 2016. 'The Compliance and Implementation Mechanism of the Paris Agreement.' *Review of European, Comparative & International Environmental Law* 25 (2): 161–73.

Voigt, Christina, and Felipe Ferreira. 2016. '"Dynamic Differentiation": The Principles of CBDR-RC, Progression and Highest Possible Ambition in the Paris Agreement.' *Transnational Environmental Law* 5 (2): 285–303.

von Lucke, Franziskus. 2017. 'Climate Justice from Kyoto to Paris.' Accessed September 01, 2017. http://www.globus.uio.no/resources/global-justice-blog/von-lucke-from-kyoto-to-paris.html.

von Lucke, Franziskus. 2020a. 'Just Power Europe or Overpowering the South? Exploring the EU's Relations with Developing Countries in the Climate Field.' *GLOBUS Research Papers* 11/2020.

von Lucke, Franziskus. 2020b. *The Securitisation of Climate Change and the Governmentalisation of Security.* Basingstoke: Palgrave Macmillan.

von Lucke, Franziskus, Thomas Diez, and Zehra Wellmann. 2014. 'What's at Stake in Securitising Climate Change? Towards a Differentiated Approach.' *Geopolitics* 19 (4): 857–84.

Wæver, Ole. 1995. 'Securitization and Desecuritization.' In *On Security*, edited by Ronnie D. Lipschutz, 46–86. New York: Columbia University Press.

Walker, Hayley. 2018. 'Mutual Recognition in Global Negotiations: The Case of the 2015 Paris Climate Summit.' *GLOBUS Research Paper* 8/2018.

Walsh, Sean, Huifang Tian, John Whalley, and Manmohan Agarwal. 2011. 'China and India's Participation in Global Climate Negotiations.' *International Environmental Agreements: Politics, Law and Economics* 11 (3): 261–73.

Warlenius, Rikard. 2018. 'Decolonizing the Atmosphere: The Climate Justice Movement on Climate Debt.' *The Journal of Environment & Development* 27 (2): 131–55.

Weidner, Helmut, and Lutz Mez. 2008. 'German Climate Change Policy: A Success Story with Some Flaws.' *The Journal of Environment & Development* 17 (4): 356–78.

Weinert, Matthew S. 2011. 'Reframing the Pluralist-Solidarist Debate.' *Millennium – Journal of International Studies* 40 (1): 21–41.

Weischer, Lutz, Jennifer Morgan, and Milap Patel. 2012. 'Climate Clubs: Can Small Groups of Countries Make a Big Difference in Addressing Climate Change?' *Review of European Community & International Environmental Law* 21 (3): 177–92.

Werksman, Jacob, Jürgen Lefevere, and Artur Runge-Metzger. 2015. 'The EU and International Climate Policy.' In *EU Climate Policy Explained*, edited by Jos Delbeke and Peter Vis, 94–112. London: Routledge.

Wheeler, Nicholas J. 1992. 'Pluralist or Solidarist Conceptions of International Society: Bull and Vincent on Humanitarian Intervention.' *Millennium – Journal of International Studies* 21 (3): 463–87.

Wheeler, Nicholas J. 2000. *Saving Strangers: Humanitarian Intervention in International Society.* Oxford: Oxford University Press.

Wheeler, Nicholas J., and Timothy Dunne. 1996. 'Hedley Bull's Pluralism of the Intellect and Solidarism of the Will.' *International Affairs* 72 (1): 91–107.

Whitman, Richard G., ed. 2011. *Normative Power Europe: Empirical and Theoretical Perspectives.* Palgrave Studies in European Union Politics. Basingstoke: Palgrave Macmillan.

Wiener, Antje. 2007. 'Contested Meanings of Norms: A Research Framework.' *Comparative European Politics* 5 (1): 1–17.

Wiener, Antje. 2017. 'A Theory of Contestation—A Concise Summary of Its Argument and Concepts.' *Polity* 49 (1): 109–25.

Williams, John. 2005. 'Pluralism, Solidarism and the Emergence of World Society in English School Theory.' *International Relations* 19 (1): 19–38.

Williams, Mariama. 2019. '24th Conference of the Parties of the UNFCCC: The US COP?' *South Centre: South Views*, February 8. 175. Accessed August 20, 2020. https://us5.campaign-archive.com/?u=fa9cf38799136b5660f367ba6&id=b197369884.

Wu, Fuzuo. 2012. 'Sino–Indian Climate Cooperation: Implications for the International Climate Change Regime.' *Journal of Contemporary China* 21 (77): 827–43.

Wurzel, Rüdiger, and James Connelly. 2011a. 'Introduction: European Union Political Leadership in International Climate Change Politics.' In *The European Union as a Leader in International Climate Change Politics*, edited by Rüdiger Wurzel and James Connelly, 3–20. Routledge/UACES Contemporary European Studies 15. London: Routledge.

Wurzel, Rüdiger, and James Connelly, eds. 2011b. *The European Union as a Leader in International Climate Change Politics.* Routledge/UACES contemporary European studies 15. London: Routledge.

Wurzel, Rüdiger K. W., James Connelly, and Duncan Liefferink, eds. 2017. *The European Union in International Climate Change Politics: Still Taking a Lead?* Abingdon: Routledge.

Yamin, Farhana. 2010. 'The Role of the EU in Climate Negotiations.' In *Climate Change and European Leadership: A Sustainable Role for Europe?*, edited by Joyeeta Gupta and Michael Grubb, 47–66. Environment and Policy v.27. Dordrecht: Springer.

Yamin, Farhana, and Joanna Depledge. 2004. *The International Climate Change Regime: A Guide to Rules, Institutions and Procedures.* Cambridge: Cambridge University Press.

Young, Iris M. 2011. *Justice and the Politics of Difference.* Princeton, NJ: Princeton University Press.

Zellentin, Alexa. 2015a. 'Climate Justice, Small Island Developing States & Cultural Loss.' *Climatic Change* 133 (3): 491–98.

Zellentin, Alexa. 2015b. 'How to Do Climate Justice.' In *Current Controversies in Political Philosophy*, edited by Thom Brooks, 121–37. Current Controversies in Philosophy. New York: Routledge.

Zhang, Yongjin. 2016. 'China and Liberal Hierarchies in Global International Society: Power and Negotiation for Normative Change.' *International Affairs* 92 (4): 795–816.

Zuma, Jacob. 2009. 'Address by President Jacob Zuma at UN Climate Change Conference on 18 December.' Accessed August 20, 2020. http://www.thepresidency.gov.za/speeches/address-president-jacob-zuma-un-climate-change-conference-copenhagen.

Zuma, Jacob. 2015. 'South Africa's Position on Climate Change Ahead of UNFCCC COP 21 Summit on 12 November.' Accessed August 20, 2020. http://www.thepresidency.gov.za/content/south-africa%E2%80%99s-position-climate-change-ahead-unfccc-cop-21-summit.

Index

Note: **Bold** page numbers refer to tables; *italic* page numbers refer to figures and Page numbers followed by "n" refer to notes.

For Product Safety Concerns and Information please contact our EU
representative GPSR@taylorandfrancis.com
Taylor & Francis Verlag GmbH, Kaufingerstraße 24, 80331 München, Germany

www.ingramcontent.com/pod-product-compliance
Lightning Source LLC
LaVergne TN
LVHW021130300926
849160LV00010B/939
* 9 7 8 0 3 6 7 5 1 1 7 2 2 *